NUREG-1801, Vol.1, Rev. 1

Generic Aging Lessons Learned (GALL) Report

Summary

Manuscript Completed: September 2005
Date Published: September 2005

Division of Regulatory Improvement Programs
Office of Nuclear Reactor Regulation
U.S. Nuclear Regulatory Commission
Washington, DC 20555-0001

ABSTRACT

NUREG-1801, "Generic Aging Lessons Learned (GALL) Report," is referenced as a technical basis document in NUREG-1800, "Standard Review Plan for Review of License Renewal Applications for Nuclear Power Plants" (SRP-LR). The GALL Report identifies aging management programs (AMP), which were determined to be acceptable programs to manage the aging effects of systems, structures and components (SSC) in the scope of license renewal, as required by 10 CFR Part 54, "Requirements for Renewal of Operating Licenses for Nuclear Power Plants."

The GALL Report is split into two volumes. Volume 1 summarizes the aging management reviews that are discussed in Volume 2. Volume 2 lists generic aging management reviews (AMRs) of SSC that may be in the scope of License Renewal Applications (LRAs) and identifies GALL AMPs that are acceptable to manage the listed aging effects. Revision 1 of the GALL Report incorporates changes based on experience gained from numerous NRC staff reviews of LRAs and other insights identified by stakeholders.

If an LRA references the GALL Report as the approach used to manage aging effect(s), the NRC staff will use the GALL Report as a basis for the LRA assessment consistent with guidance specified in the SRP-LR.

TABLE OF CONTENTS

LIST OF TABLES

LIST OF CONTRIBUTORS

License Renewal and Environmental Programs Section of Division of Regulatory Improvement
Programs, Office of Nuclear Reactor Regulation

P.T. Kuo	Program Director
S. Lee	Section Chief
S. West	Section Chief
J. Zimmerman	Section Chief
J. Dozier	Team Leader
K. Chang	Mechanical Engineering
K. Cozens	Materials Engineering
G. Cranston	Reactor Systems Engineering
D. Guha	Systems Engineering
M. Heath	Mechanical Engineering
S. Hoffman	Mechanical Engineering
A. Hull	Materials Engineering
K. Hsu	Materials Engineering
M. Lintz	Mechanical Engineering
D. Merzke	Mechanical Engineering
K. Naidu	Reactor Engineering
J. Rajan	Mechanical Engineering
R. Subbaratnam	Mechanical Engineering
T. Terry	Civil Engineering
L. Tran	Electrical Engineering
P. Wen	Electrical Engineering

Office of Nuclear Reactor Regulation

T. Chan	Section Chief
S. Coffin	Section Chief
S. Jones	Section Chief
R. Jenkins	Section Chief
L. Lund	Section Chief
R. Karas	Section Chief
K. Manoly	Section Chief
M. Mitchell	Section Chief
J. Nakoski	Section Chief
D. Terao	Section Chief
S. Weerakkody	Section Chief
H. Ashar	Structural Engineering
S. Bailey	Mechanical Engineering
T. Cheng	Structural Engineering
R. Davis	Materials Engineering
B. Elliot	Materials Engineering
J. Fair	Mechanical Engineering
G. Georgiev	Materials Engineering

A. Keim	Materials Engineering
N. Iqbal	Fire Protection Engineering
D. Jeng	Structural Engineering
K. Karwoski	Materials Engineering
C. Lauron	Chemical Engineering
L. Lois	Reactor Systems Engineering
Y. Li	Mechanical Engineering
R. McNally	Mechanical Engineering
J. Medoff	Materials Engineering
D. Nguyen	Electrical Engineering
A. Pal	Electrical Engineering
K. Parczewski	Chemical Engineering
J. Strnisha	Mechanical Engineering
P. Shemanski	Electrical Engineering

Office of Nuclear Regulatory Research

A. Hiser	Section Chief
J. Vora	Team Leader
J. Davis	Materials Engineering
P. Kang	Electrical Engineering

Parallax, Inc

A. Baione	Team Leader
M. Bowman	Mechanical Engineering
D. Jones	Programming
K. Larsen	Technical Editing
E. Patel	Mechanical Engineering
R. Wells	License Engineering

ABBREVIATIONS

ADS	automatic depressurization system
AFW	auxiliary feedwater
AMP	aging management program
ASME	American Society of Mechanical Engineers
B&W	Babcock & Wilcox
BWR	boiling water reactor
BWRVIP	boiling water reactor vessel internals project
CASS	cast austenitic stainless steel
CE	Combustion Engineering
CEA	control element assembly
CFR	Code of Federal Regulations
CFS	core flood system
CLB	current licensing basis
CRD	control rod drive
CRGT	control rod guide tube
CS	carbon steel
CVCS	chemical and volume control system
DHR	decay heat removal
DSCSS	drywell and suppression chamber spray system
ECCS	emergency core cooling system
EDG	emergency diesel generator
EQ	environmental qualification
FW	feedwater
GALL	generic aging lessons learned
HP	high pressure
HPCI	high-pressure coolant injection
HPCS	high-pressure core spray
HPSI	high-pressure safety injection
HVAC	heating, ventilation, and air conditioning
IASCC	irradiation-assisted stress corrosion cracking
IGA	intergranular attack
IGSCC	intergranular stress corrosion cracking
IR	insulation resistance
IRM	intermediate range monitor
ISI	inservice inspection
LER	licensee event report
LG	lower grid

ABBREVIATIONS (continued)

LP	low pressure
LPCI	low-pressure coolant injection
LPCS	low-pressure core spray
LPRM	low-power range monitor
LPSI	low-pressure safety injection
MIC	microbiologically influenced corrosion
MSR	moisture separator/reheater
NEI	Nuclear Energy Institute
NPAR	Nuclear Plant Aging Research
NPS	nominal pipe size
NRC	Nuclear Regulatory Commission
NSSS	nuclear steam supply system
NUMARC	Nuclear Management and Resources Council
ODSCC	outside diameter stress corrosion cracking
PWR	pressurized water reactor
PWSCC	primary water stress corrosion cracking
QA	quality assurance
RCCA	rod control cluster assembly
RCIC	reactor core isolation cooling
RCP	reactor coolant pump
RCPB	reactor coolant pressure boundary
RCS	reactor coolant system
RG	Regulatory Guide
RHR	residual heat removal
RWC	reactor water cleanup
RWT	refueling water tank
SBO	station blackout
SC	suppression chamber
SCC	stress corrosion cracking
SDC	shutdown cooling
SFP	spent fuel pool
SG	steam generator
SLC	standby liquid control
SRM	source range monitor
SRM	staff requirement memorandum
SRP-LR	Standard Review Plan for License Renewal
TLAA	time-limited aging analysis
UCS	Union of Concerned Scientists
UV	ultraviolet

INTRODUCTION

NUREG-1801, "Generic Aging Lessons Learned (GALL) Report," is referenced as a technical basis document in NUREG-1800, "Standard Review Plan for Review of License Renewal Applications for Nuclear Power Plants" (SRP-LR). The GALL Report identifies aging management programs (AMP) that were determined to be acceptable to manage aging effects of systems, structures and components (SSC) in the scope of license renewal, as required by 10 CFR Part 54, "Requirements for Renewal of Operating Licenses for Nuclear Power Plants."

The GALL Report is comprised of two volumes. Volume 1 summarizes the aging management reviews that are discussed in Volume 2. Volume 2 lists generic aging management reviews (AMRs) of SSCs that may be in the scope of license renewal applications (LRAs) and identifies GALL AMPs that are acceptable to manage the aging effects.

If an LRA references the GALL Report as the approach used to manage aging effect(s), the NRC staff will use the GALL Report as a basis for the LRA assessment consistent with guidance specified in the SRP-LR.

BACKGROUND

Revision 0 of the GALL Report

By letter dated March 3, 1999, the Nuclear Energy Institute (NEI) documented the industry's views on how existing plant programs and activities should be credited for license renewal. The issue can be summarized as follows: To what extent should the staff review existing programs relied on for license renewal in determining whether an applicant has demonstrated reasonable assurance that such programs will be effective in managing the effects of aging on the functionality of structures and components during the period of extended operation? In a staff paper, SECY-99-148, "Credit for Existing Programs for License Renewal," dated June 3, 1999, the staff described options for crediting existing programs and recommended one option that the staff believed would improve the efficiency of the license renewal process.

By staff requirements memorandum (SRM), dated August 27, 1999, the Commission approved the staff's recommendation and directed the staff to focus the staff review guidance in the Standard Review Plan for License Renewal (SRP-LR) on areas where existing programs should be augmented for license renewal. The staff would develop a "Generic Aging Lessons Learned (GALL)" report to document the staff's evaluation of generic existing programs. The GALL Report would document the staff's basis for determining which existing programs are adequate without modification and which existing programs should be augmented for license renewal. The GALL Report would be referenced in the SRP-LR as a basis for determining the adequacy of existing programs.

This report builds on a previous report, NUREG/CR-6490, "Nuclear Power Plant Generic Aging Lessons Learned (GALL)," which is a systematic compilation of plant aging information. This report extends the information in NUREG/CR-6490 to provide an evaluation of the adequacy of aging management programs for license renewal. The NUREG/CR-6490 report was based on information in over 500 documents: Nuclear Plant Aging Research (NPAR) program reports sponsored by the Office of Nuclear Regulatory Research, Nuclear Management and Resources Council (NUMARC, now NEI) industry reports addressing license renewal for major structures and components, licensee event reports (LERs), information notices, generic letters, and bulletins. The staff has also considered information contained in the reports provided by the Union of Concerned Scientists (UCS) in a letter dated May 5, 2000.

Following the general format of NUREG-0800 for major plant sections except for refueling water, chilled water, residual heat removal, condenser circulating water, and condensate storage system in pressurized water reactor (PWR) and boiling water reactor (BWR) power plants, the staff has reviewed the aging effects on components and structures, identified the relevant existing programs, and evaluated program attributes to manage aging effects for license renewal. This report was prepared with the technical assistance of Argonne National Laboratory and Brookhaven National Laboratory. As directed in the SRM, this report has the benefit of the experience of the staff members who conducted the review of the initial license renewal applications. Also, as directed in the SRM, the staff has sought stakeholders' participation in the development of this report. The staff held many public meetings and workshops to solicit input from the public. The staff also requested comments from the public on the draft improved license renewal guidance documents, including the GALL Report, in the Federal Register Notice, Vol. 65, No. 170, August 31, 2000. The staff's analysis of stakeholder comments is documented in NUREG-1739. These documents can be found on-line at: http://www.nrc.gov/reading-rm/doc-collections/.

Revision 1 of the GALL Report

The GALL Report has been referenced in numerous license renewal applications (LRA) as a basis for aging management reviews to satisfy the regulatory criteria contained in 10 CFR Part 54, "Requirements for Renewal of Operating Licenses for Nuclear Power Plants," Section 54.21, "Contents of application – technical information." Based on lessons learned from these reviews, and other public input, including industry comments, the NRC staff proposed changes to the GALL Report to make the GALL Report more efficient. A preliminary version of Revision 1 of the GALL Report was posted on the NRC public web page on September 30, 2004. The draft revisions of GALL Vol. 1 and Vol. 2 were further refined and issued for public comment on January 31, 2005. In addition, the staff also held public meetings with stakeholders to facilitate dialog and to discuss comments. The staff subsequently took into consideration comments received (see NUREG-1832) and incorporated its dispositions into the September 2005 version of the GALL Report.

OVERVIEW OF THE GALL REPORT EVALUATION PROCESS

The results of the GALL effort are presented in a table format in the GALL Report, Volume 2. The table column headings are: Item, Structure and/or Component; Material, Environment; Aging Effect/Mechanism; Aging Management Program (AMP); and Further Evaluation. The staff's evaluation of the adequacy of each generic aging management program in managing certain aging effects for particular structures and components is based on its review of the following 10 program elements in each aging management program:

AMP Element	Description
1. Scope of the program	The scope of the program should include the specific structures and components subject to an aging management review.
2. Preventive actions	Preventive actions should mitigate or prevent the applicable aging effects.
3. Parameters monitored or inspected	Parameters monitored or inspected should be linked to the effects of aging on the intended functions of the particular

AMP Element	Description
	structure and component.
4. Detection of aging effects	Detection of aging effects should occur before there is a loss of any structure and component intended function. This includes aspects such as method or technique (i.e., visual, volumetric, surface inspection), frequency, sample size, data collection and timing of new/one-time inspections to ensure timely detection of aging effects.
5. Monitoring and trending	Monitoring and trending should provide for prediction of the extent of the effects of aging and timely corrective or mitigative actions.
6. Acceptance criteria	Acceptance criteria, against which the need for corrective action will be evaluated, should ensure that the particular structure and component intended functions are maintained under all current licensing basis (CLB) design conditions during the period of extended operation.
7. Corrective actions	Corrective actions, including root cause determination and prevention of recurrence, should be timely.
8. Confirmation process	The confirmation process should ensure that preventive actions are adequate and appropriate corrective actions have been completed and are effective.
9. Administrative controls	Administrative controls should provide a formal review and approval process.
10. Operating experience	Operating experience involving the aging management program, including past corrective actions resulting in program enhancements or additional programs, should provide objective evidence to support a determination that the effects of aging will be adequately managed so that the structure and component intended functions will be maintained during the period of extended operation.

If, on the basis of its evaluation, the staff determined that a program is adequate to manage certain aging effects for a particular structure or component without change, the "Further Evaluation" entry would indicate that no further evaluation is recommended for license renewal.

Chapter XI of the GALL Report, Volume 2, contains the staff's evaluation of generic aging management programs that are relied on in the GALL Report, such as the ASME Section XI inservice inspection, water chemistry, or structures monitoring program.

APPLICATION OF THE GALL REPORT

The GALL Report is a technical basis document to the SRP-LR, which provides the staff with guidance in reviewing a license renewal application. The GALL Report should be treated in the same manner as an approved topical report that is generically applicable. An applicant may reference the GALL Report in a license renewal application to demonstrate that the programs at the applicant's facility correspond to those reviewed and approved in the GALL Report.

If an applicant takes credit for a program in GALL, it is incumbent on the applicant to ensure that the plant program contains all the elements of the referenced GALL program. In addition, the conditions at the plant must be bounded by the conditions for which the GALL program was evaluated. The above verifications must be documented on-site in an auditable form. The applicant must include a certification in the license renewal application that the verifications have been completed.

The GALL Report contains one acceptable way to manage aging effects for license renewal. An applicant may propose alternatives for staff review in its plant-specific license renewal application. Use of the GALL Report is not required, but its use should facilitate both preparation of a license renewal application by an applicant and timely, uniform review by the NRC staff.

In addition, the GALL Report does not address scoping of structures and components for license renewal. Scoping is plant specific, and the results depend on the plant design and current licensing basis. The inclusion of a certain structure or component in the GALL Report does not mean that this particular structure or component is within the scope of license renewal for all plants. Conversely, the omission of a certain structure or component in the GALL Report does not mean that this particular structure or component is not within the scope of license renewal for any plants.

The GALL Report contains an evaluation of a large number of structures and components that may be in the scope of a typical LRA. The evaluation results documented in the GALL Report indicate that many existing, typical generic aging management programs are adequate to manage aging effects for particular structures or components for license renewal without change. The GALL Report also contains recommendations on specific areas for which generic existing programs should be augmented (require further evaluation) for license renewal and documents the technical basis for each such determination. In addition, the GALL Report identifies certain SSCs that may or may not be subject to particular aging effects, and for which industry groups are developing generic aging management programs or investigating whether aging management is warranted. To the extent the ultimate generic resolution of such an issue will need NRC review and approval for plant-specific implementation, as indicated in a plant-specific FSAR supplement, and reflected in the SER associated with a particular LR application, an amendment pursuant to 10 CFR 50.90 will be necessary.

In the GALL Report, Volume 1, Tables 1 through 6 are summaries of the aging management review. These tables contain the same information as Tables 3.1-1 to 3.6-1, respectively, in the SRP-LR. These tables also include additional seventh and eighth columns that identify the related generic item and unique item associated with each structure and/or component (i.e., each row in the AMR tables contained in Volume 2 of the GALL Report). A locator for the plant systems evaluated in Volume 2 is also provided in the Appendix of Volume 1.

The Appendix of Volume 2 of the GALL Report addresses quality assurance (QA) for aging management programs. Those aspects of the aging management review process that affect the quality of safety-related structures, systems, and components are subject to the QA requirements of Appendix B to 10 CFR Part 50. For nonsafety-related structures and components subject to an aging management review, the existing 10 CFR Part 50, Appendix B, QA program may be used by an applicant to address the elements of the corrective actions, confirmation process, and administrative controls for an aging management program for license renewal.

The GALL Report provides a technical basis for crediting existing plant programs and recommending areas for program augmentation and further evaluation. The incorporation of the GALL Report information into the SRP-LR, as directed by the Commission, should improve the efficiency of the license renewal process and better focus staff resources.

Table Column Headings

The following describes the information presented in each column of Tables 1 through 6 contained in Volume 1 of this report. These tables present the relationship between the SRP-LR lines, the unique AMR line-item identifier (unique item) and the chapter-specific generic item that can be referenced repeatedly within a given chapter of GALL Vol. 2.

Column Heading	Description
ID	A unique row identifier. This identifier is useful in matching the row with the row in the corresponding 3.X-1 Table in the SRP-LR (where the "X" represents the chapter number within the SRP-LR). Thus, the Table 1 row labeled ID 1 in GALL Vol. 1 represents the same information contained in the row labeled ID 1 in Table 3.1-1 of the SRP-LR.
Type	Identifies the plant design that the item applies to (i.e., BWR or PWR or both).
Component	Identifies the structure or components to which the row applies
Aging Effect/ Mechanism	Identifies the applicable aging effect and mechanism(s). See Chapter IX of Volume 2 for more information.
Aging Management Programs	Identifies the time limited aging analysis or aging management program found acceptable for properly managing the affects of aging. See Chapter X and XI of Volume 2.
Further Evaluation Recommended	Identifies whether further evaluation is required, and references the section of the SRP-LR that provides further information on this evaluation.
Related Generic Item	Identifies the item number in Volume 2, Chapters II through VIII presenting the detailed information summarized by this row. This chapter-specific generic identifier is used in the AMR subsystem rows and can appear multiple times within a chapter.
Unique Item	The unique item is an AMR line-item identifier which is coded to indicate the chapter, AMR subsystem and unique row number within GALL Volume 2 (i.e., VIII.B1-1 is the first row in the steam and power conversion system, main steam system table, row 1).

Table 1. Summary of Aging Management Programs for the Reactor Coolant System Evaluated in Chapter IV of the GALL Report

ID	Type	Component	Aging Effect/Mechanism	Aging Management Programs	Further Evaluation Recommended	Related Generic Item	Unique Item
1	BWR	Steel pressure vessel support skirt and attachment welds	Cumulative fatigue damage	TLAA, evaluated in accordance with 10 CFR 54.21(c)	Yes, TLAA	R-70	IV.A1-6 IV.A2-20
2	BWR	Steel; stainless steel; steel with nickel-alloy or stainless steel cladding; nickel-alloy reactor vessel components: flanges; nozzles; penetrations; safe ends; thermal sleeves; vessel shells, heads and welds	Cumulative fatigue damage	TLAA, evaluated in accordance with 10 CFR 54.21(c) and environmental effects are to be addressed for Class 1 components	Yes, TLAA	R-04	IV.A1-7
3	BWR	Steel; stainless steel; steel with nickel-alloy or stainless steel cladding; nickel-alloy reactor coolant pressure boundary piping, piping components, and piping elements exposed to reactor coolant	Cumulative fatigue damage	TLAA, evaluated in accordance with 10 CFR 54.21(c) and environmental effects are to be addressed for Class 1 components	Yes, TLAA	R-220	IV.C1-15
4	BWR	Steel pump and valve closure bolting	Cumulative fatigue damage	TLAA, evaluated in accordance with 10 CFR 54.21(c) check Code limits for allowable cycles (less than 7000 cycles) of thermal stress range	Yes, TLAA	R-28	IV.C1-11
5	BWR/ PWR	Stainless steel and nickel alloy reactor vessel internals components	Cumulative fatigue damage	TLAA, evaluated in accordance with 10 CFR 54.21(c)	Yes, TLAA	R-53	IV.B1-14 IV.B2-31 IV.B3-24 IV.B4-37

Table 1. Summary of Aging Management Programs for the Reactor Coolant System Evaluated in Chapter IV of the GALL Report

ID	Type	Component	Aging Effect/Mechanism	Aging Management Programs	Further Evaluation Recommended	Related Generic Item	Unique Item
6	PWR	Nickel Alloy tubes and sleeves in a reactor coolant and secondary feedwater/steam environment	Cumulative fatigue damage	TLAA, evaluated in accordance with 10 CFR 54.21(c)	Yes, TLAA	R-46	IV.D1-21 IV.D2-15
7	PWR	Steel and stainless steel reactor coolant pressure boundary closure bolting, head closure studs, support skirts and attachment welds, pressurizer relief tank components, steam generator components, piping and components external surfaces and bolting	Cumulative fatigue damage	TLAA, evaluated in accordance with 10 CFR 54.21(c)	Yes, TLAA	R-13 R-18 R-33 R-73	IV.C2-23 IV.C2-10 IV.D1-11 IV.D2-10 IV.A2-4
8	PWR	Steel; stainless steel; and nickel-alloy reactor coolant pressure boundary piping, piping components, piping elements; flanges; nozzles and safe ends; pressurizer vessel shell heads and welds; heater sheaths and sleeves; penetrations; and thermal sleeves	Cumulative fatigue damage	TLAA, evaluated in accordance with 10 CFR 54.21(c) and environmental effects are to be addressed for Class 1 components	Yes, TLAA	R-223	IV.C2-25
9	PWR	Steel; stainless steel; steel with nickel-alloy or stainless steel cladding; nickel-alloy reactor vessel components: flanges; nozzles; penetrations; pressure housings; safe ends; thermal sleeves; vessel shells, heads and welds	Cumulative fatigue damage	TLAA, evaluated in accordance with 10 CFR 54.21(c) and environmental effects are to be addressed for Class 1 components	Yes, TLAA	R-219	IV.A2-21

Table 1. Summary of Aging Management Programs for the Reactor Coolant System Evaluated in Chapter IV of the GALL Report

ID	Type	Component	Aging Effect/Mechanism	Aging Management Programs	Further Evaluation Recommended	Related Generic Item	Unique Item
10	PWR	Steel; stainless steel; steel with nickel-alloy or stainless steel cladding; nickel-alloy steam generator components (flanges; penetrations; nozzles; safe ends, lower heads and welds)	Cumulative fatigue damage	TLAA, evaluated in accordance with 10 CFR 54.21(c) and environmental effects are to be addressed for Class 1 components	Yes, TLAA	R-221 R-222	IV.D1-8 IV.D2-3
11	BWR	Steel top head enclosure (without cladding) top head nozzles (vent, top head spray or RCIC, and spare) exposed to reactor coolant	Loss of material due to general, pitting and crevice corrosion	Water Chemistry and One-Time Inspection	Yes, detection of aging effects is to be evaluated	R-59	IV.A1-11
12	PWR	Steel steam generator shell assembly exposed to secondary feedwater and steam	Loss of material due to general, pitting and crevice corrosion	Water Chemistry and One-Time Inspection	Yes, detection of aging effects is to be evaluated	R-224	IV.D2-8
13	BWR	Steel and stainless steel isolation condenser components exposed to reactor coolant	Loss of material due to general (steel only), pitting and crevice corrosion	Water Chemistry and One-Time Inspection	Yes, detection of aging effects is to be evaluated	R-16	IV.C1-6
14	BWR	Stainless steel, nickel-alloy, and steel with nickel-alloy or stainless steel cladding reactor vessel flanges, nozzles, penetrations, safe ends, vessel shells, heads and welds	Loss of material due to pitting and crevice corrosion	Water Chemistry and One-Time Inspection	Yes, detection of aging effects is to be evaluated	RP-25	IV.A1-8

Table 1. Summary of Aging Management Programs for the Reactor Coolant System Evaluated in Chapter IV of the GALL Report

ID	Type	Component	Aging Effect/Mechanism	Aging Management Programs	Further Evaluation Recommended	Related Generic Item	Unique Item
15	BWR	Stainless steel; steel with nickel-alloy or stainless steel cladding; and nickel-alloy reactor coolant pressure boundary components exposed to reactor coolant	Loss of material due to pitting and crevice corrosion	Water Chemistry and One-Time Inspection	Yes, detection of aging effects is to be evaluated	RP-27	IV.C1-14
16	PWR	Steel steam generator upper and lower shell and transition cone exposed to secondary feedwater and steam	Loss of material due to general, pitting and crevice corrosion	Inservice Inspection (IWB, IWC, and IWD), and Water Chemistry and, for Westinghouse Model 44 and 51 S/G, if general and pitting corrosion of the shell is known to exist, additional inspection procedures are to be developed.	Yes, detection of aging effects is to be evaluated	R-34	IV.D1-12
17	BWR/ PWR	Steel (with or without stainless steel cladding) reactor vessel beltline shell, nozzles, and welds	Loss of fracture toughness due to neutron irradiation embrittlement	TLAA, evaluated in accordance with Appendix G of 10 CFR 50 and RG 1.99. The applicant may choose to demonstrate that the materials of the nozzles are not controlling for the TLAA evaluations.	Yes, TLAA	R-62 R-67 R-81 R-84	IV.A1-13 IV.A1-4 IV.A2-16 IV.A2-23

Table 1. Summary of Aging Management Programs for the Reactor Coolant System Evaluated in Chapter IV of the GALL Report

ID	Type	Component	Aging Effect/Mechanism	Aging Management Programs	Further Evaluation Recommended	Related Generic Item	Unique Item
18	BWR/PWR	Steel (with or without stainless steel cladding) reactor vessel beltline shell, nozzles, and welds; safety injection nozzles	Loss of fracture toughness due to neutron irradiation embrittlement	Reactor Vessel Surveillance	Yes, plant specific	R-63 R-82 R-86	IV.A1-14 IV.A2-17 IV.A2-24
19	BWR	Stainless steel and nickel alloy top head enclosure vessel flange leak detection line	Cracking due to stress corrosion cracking and intergranular stress corrosion cracking	A plant-specific aging management program is to be evaluated because existing programs may not be capable of mitigating or detecting crack initiation and growth due to SCC in the vessel flange leak detection line.	Yes, plant specific	R-61	IV.A1-10
20	BWR	Stainless steel isolation condenser components exposed to reactor coolant	Cracking due to stress corrosion cracking and intergranular stress corrosion cracking	Inservice Inspection (IWB, IWC, and IWD), Water Chemistry, and plant-specific verification program	Yes, detection of aging effects is to be evaluated	R-15	IV.C1-4
21	PWR	Reactor vessel shell fabricated of SA508-Cl 2 forgings clad with stainless steel using a high-heat-input welding process	Crack growth due to cyclic loading	TLAA	Yes, TLAA	R-85	IV.A2-22

Table 1. Summary of Aging Management Programs for the Reactor Coolant System Evaluated in Chapter IV of the GALL Report

ID	Type	Component	Aging Effect/Mechanism	Aging Management Programs	Further Evaluation Recommended	Related Generic Item	Unique Item
22	PWR	Stainless steel and nickel alloy reactor vessel internals components exposed to reactor coolant and neutron flux	Loss of fracture toughness due to neutron irradiation embrittlement, void swelling	FSAR supplement commitment to (1) participate in industry RVI aging programs (2) implement applicable results (3) submit for NRC approval > 24 months before the extended period an RVI inspection plan based on industry recommendation.	No, but licensee commitment to be confirmed	R-122 R-127 R-128 R-132 R-135 R-141 R-157 R-161 R-164 R-169 R-178 R-188 R-196 R-205 R-212 R-216	IV.B2-9 IV.B2-3 IV.B2-6 IV.B4-1 IV.B2-18 IV.B2-17 IV.B2-22 IV.B3-16 IV.B3-12 IV.B3-10 IV.B3-20 IV.B4-46 IV.B4-16 IV.B4-12 IV.B4-31 IV.B4-24 IV.B4-41
23	PWR	Stainless steel reactor vessel closure head flange leak detection line and bottom-mounted instrument guide tubes	Cracking due to stress corrosion cracking	A plant-specific aging management program is to be evaluated.	Yes, plant specific	R-74 RP-13	IV.A2-5 IV.A2-1
24	PWR	Class 1 cast austenitic stainless steel piping, piping components, and piping elements exposed to reactor coolant	Cracking due to stress corrosion cracking	Water Chemistry and, for CASS components that do not meet the NUREG-0313 guidelines, a plant specific aging management program	Yes, plant specific	R-05	IV.C2-3

Table 1. Summary of Aging Management Programs for the Reactor Coolant System Evaluated in Chapter IV of the GALL Report

ID	Type	Component	Aging Effect/Mechanism	Aging Management Programs	Further Evaluation Recommended	Related Generic Item	Unique Item
25	BWR	Stainless steel jet pump sensing line	Cracking due to cyclic loading	A plant-specific aging management program is to be evaluated.	Yes, plant specific	R-102	IV.B1-12
26	BWR	Steel and stainless steel isolation condenser components exposed to reactor coolant	Cracking due to cyclic loading	Inservice Inspection (IWB, IWC, and IWD) and plant-specific verification program	Yes, detection of aging effects is to be evaluated	R-225	IV.C1-5
27	PWR	Stainless steel and nickel alloy reactor vessel internals screws, bolts, tie rods, and hold-down springs	Loss of preload due to stress relaxation	FSAR supplement commitment to (1) participate in industry RVI aging programs (2) implement applicable results (3) submit for NRC approval > 24 months before the extended period an RVI inspection plan based on industry recommendation.	No, but licensee commitment to be confirmed	R-108 R-114 R-129 R-136 R-137 R-154 R-165 R-184 R-192 R-197 R-201 R-207 R-213	IV.B2-33 IV.B2-38 IV.B2-5 IV.B2-25 IV.B2-14 IV.B3-6 IV.B3-7 IV.B4-6 IV.B4-19 IV.B4-14 IV.B4-9 IV.B4-33 IV.B4-26
28	PWR	Steel steam generator feedwater impingement plate and support exposed to secondary feedwater	Loss of material due to erosion	A plant-specific aging management program is to be evaluated.	Yes, plant specific	R-39	IV.D1-13
29	BWR	Stainless steel steam dryers exposed to reactor coolant	Cracking due to flow-induced vibration	A plant-specific aging management program is to be evaluated.	Yes, plant specific	RP-18	IV.B1-16

Table 1. Summary of Aging Management Programs for the Reactor Coolant System Evaluated in Chapter IV of the GALL Report

ID	Type	Component	Aging Effect/Mechanism	Aging Management Programs	Further Evaluation Recommended	Related Generic Item	Unique Item
30	PWR	Stainless steel reactor vessel internals components (e.g., Upper internals assembly, RCCA guide tube assemblies, Baffle/former assembly, Lower internal assembly, shroud assemblies, Plenum cover and plenum cylinder, Upper grid assembly, Control rod guide tube (CRGT) assembly, Core support shield assembly, Core barrel assembly, Lower grid assembly, Flow distributor assembly, Thermal shield, Instrumentation support structures)	Cracking due to stress corrosion cracking, irradiation-assisted stress corrosion cracking	Water Chemistry and FSAR supplement commitment to (1) participate in industry RVI aging programs (2) implement applicable results (3) submit for NRC approval > 24 months before the extended period an RVI inspection plan based on industry recommendation.	No, but licensee commitment needs to be confirmed	R-106 R-109 R-116 R-120 R-123 R-125 R-138 R-143 R-146 R-149 R-155 R-159 R-166 R-172 R-173 R-175 R-176 R-180 R-181 R-185 R-193 R-202 R-209 R-214	IV.B2-42 IV.B2-36 IV.B2-30 IV.B2-8 IV.B2-2 IV.B2-10 IV.B4-7 IV.B2-24 IV.B2-12 IV.B3-28 IV.B3-2 IV.B3-15 IV.B3-11 IV.B3-21 IV.B4-34 IV.B4-36 IV.B4-44 IV.B4-43 IV.B4-2 IV.B4-5 IV.B4-18 IV.B4-10 IV.B4-29 IV.B4-22 IV.B4-40

Table 1.   Summary of Aging Management Programs for the Reactor Coolant System Evaluated in Chapter IV of the GALL Report

ID	Type	Component	Aging Effect/Mechanism	Aging Management Programs	Further Evaluation Recommended	Related Generic Item	Unique Item
31	PWR	Nickel alloy and steel with nickel-alloy cladding piping, piping component, piping elements, penetrations, nozzles, safe ends, and welds (other than reactor vessel head); pressurizer heater sheaths, sleeves, diaphragm plate, manways and flanges; core support pads/core guide lugs	Cracking due to primary water stress corrosion cracking	Inservice Inspection (IWB, IWC, and IWD) and Water Chemistry and FSAR supp commitment to implement applicable plant commitments to (1) NRC Orders, Bulletins, and Generic Letters associated with nickel alloys and (2) staff-accepted industry guidelines.	No, but licensee commitment needs to be confirmed	R-01 R-06 R-88 R-89 RP-22 RP-31	IV.D1-4 IV.D2-2 IV.C2-21 IV.A2-12 IV.A2-19 IV.C2-24 IV.C2-13
32	PWR	Steel steam generator feedwater inlet ring and supports	Wall thinning due to flow-accelerated corrosion	A plant-specific aging management program is to be evaluated.	Yes, plant specific	R-51	IV.D1-26

Table 1. Summary of Aging Management Programs for the Reactor Coolant System Evaluated in Chapter IV of the GALL Report

ID	Type	Component	Aging Effect/Mechanism	Aging Management Programs	Further Evaluation Recommended	Related Generic Item	Unique Item
33	PWR	Stainless steel and nickel alloy reactor vessel internals components	Changes in dimensions due to void swelling	FSAR supplement commitment to (1) participate in industry RVI aging programs (2) implement applicable results (3) submit for NRC approval > 24 months before the extended period an RVI inspection plan based on industry recommendation.	No, but licensee commitment to be confirmed	R-107 R-110 R-113 R-117 R-119 R-121 R-124 R-126 R-131 R-134 R-139 R-144 R-147 R-151 R-158 R-160 R-163 R-168 R-174 R-177 R-182 R-187 R-195 R-199 R-204 R-211 R-215	IV.B2-41 IV.B2-35 IV.B2-39 IV.B2-29 IV.B2-27 IV.B2-7 IV.B2-1 IV.B2-4 IV.B2-19 IV.B2-15 IV.B2-23 IV.B2-11 IV.B3-27 IV.B3-4 IV.B3-14 IV.B3-13 IV.B3-8 IV.B3-19 IV.B4-35 IV.B4-45 IV.B4-3 IV.B4-17 IV.B4-11 IV.B4-8 IV.B4-30 IV.B4-23 IV.B4-39

Table 1. Summary of Aging Management Programs for the Reactor Coolant System Evaluated in Chapter IV of the GALL Report

ID	Type	Component	Aging Effect/Mechanism	Aging Management Programs	Further Evaluation Recommended	Related Generic Item	Unique Item
34	PWR	Stainless steel and nickel alloy reactor control rod drive head penetration pressure housings	Cracking due to stress corrosion cracking and primary water stress corrosion cracking	Inservice Inspection (IWB, IWC, and IWD) and Water Chemistry and for nickel alloy, FSAR supplement commitment to implement applicable plant commitments to (1) NRC Orders, Bulletins and Generic Letters associated with nickel alloys and (2) staff-accepted industry guidelines.	No, but licensee commitment needs to be confirmed	R-76	IV.A2-11
35	PWR	Steel with stainless steel or nickel alloy cladding primary side components; steam generator upper and lower heads, tubesheets and tube-to-tube sheet welds	Cracking due to stress corrosion cracking and primary water stress corrosion cracking	Inservice Inspection (IWB, IWC, and IWD) and Water Chemistry and for nickel alloy, FSAR supplement commitment to implement applicable plant commitments to (1) NRC Orders, Bulletins and Generic Letters associated with nickel alloys and (2) staff-accepted industry guidelines.	No, but licensee commitment needs to be confirmed	R-35	IV.D2-4

Table 1. Summary of Aging Management Programs for the Reactor Coolant System Evaluated in Chapter IV of the GALL Report

ID	Type	Component	Aging Effect/Mechanism	Aging Management Programs	Further Evaluation Recommended	Related Generic Item	Unique Item
36	PWR	Nickel alloy, stainless steel pressurizer spray head	Cracking due to stress corrosion cracking and primary water stress corrosion cracking	Water Chemistry and One-Time Inspection and, for nickel alloy welded spray heads, provide commitment in FSAR supplement to submit AMP delineating commitments to Orders, Bulletins, or Generic Letters that inspect stipulated components for cracking of wetted surfaces.	No, unless licensee commitment needs to be confirmed	R-24	IV.C2-17
37	PWR	Stainless steel and nickel alloy reactor vessel internals components (e.g., Upper internals assembly, RCCA guide tube assemblies, Lower internal assembly, CEA shroud assemblies, Core shroud assembly, Core support shield assembly, Core barrel assembly, Lower grid assembly, Flow distributor assembly)	Cracking due to stress corrosion cracking, primary water stress corrosion cracking, irradiation-assisted stress corrosion cracking	Water Chemistry and FSAR supplement commitment to (1) participate in industry RVI aging programs (2) implement applicable results (3) submit for NRC approval > 24 months before the extended period an RVI inspection plan based on industry recommendation.	No, but licensee commitment needs to be confirmed	R-112 R-118 R-130 R-133 R-150 R-162 R-167 R-186 R-194 R-203 R-210	IV.B2-40 IV.B2-28 IV.B2-20 IV.B2-16 IV.B3-5 IV.B3-9 IV.B3-23 IV.B4-20 IV.B4-13 IV.B4-32 IV.B4-25
38	BWR	Steel (with or without stainless steel cladding) control rod drive return line nozzles exposed to reactor coolant	Cracking due to cyclic loading	BWR CR Drive Return Line Nozzle	No	R-66	IV.A1-2

Table 1. Summary of Aging Management Programs for the Reactor Coolant System Evaluated in Chapter IV of the GALL Report

ID	Type	Component	Aging Effect/Mechanism	Aging Management Programs	Further Evaluation Recommended	Related Generic Item	Unique Item
39	BWR	Steel (with or without stainless steel cladding) feedwater nozzles exposed to reactor coolant	Cracking due to cyclic loading	BWR Feedwater Nozzle	No	R-65	IV.A1-3
40	BWR	Stainless steel and nickel alloy penetrations for control rod drive stub tubes instrumentation, jet pump instrument, standby liquid control, flux monitor, and drain line exposed to reactor coolant	Cracking due to stress corrosion cracking, Intergranular stress corrosion cracking, cyclic loading	BWR Penetrations and Water Chemistry	No	R-69	IV.A1-5
41	BWR	Stainless steel and nickel alloy piping, piping components, and piping elements greater than or equal to 4 NPS; nozzle safe ends and associated welds	Cracking due to stress corrosion cracking and intergranular stress corrosion cracking	BWR Stress Corrosion Cracking and Water Chemistry	No	R-20 R-21 R-68	IV.C1-9 IV.C1-8 IV.A1-1
42	BWR	Stainless steel and nickel alloy vessel shell attachment welds exposed to reactor coolant	Cracking due to stress corrosion cracking and intergranular stress corrosion cracking	BWR Vessel ID Attachment Welds and Water Chemistry	No	R-64	IV.A1-12
43	BWR	Stainless steel fuel supports and control rod drive assemblies control rod drive housing exposed to reactor coolant	Cracking due to stress corrosion cracking and intergranular stress corrosion cracking	BWR Vessel Internals and Water Chemistry	No	R-104	IV.B1-8

Table 1. Summary of Aging Management Programs for the Reactor Coolant System Evaluated in Chapter IV of the GALL Report

ID	Type	Component	Aging Effect/Mechanism	Aging Management Programs	Further Evaluation Recommended	Related Generic Item	Unique Item
44	BWR	Stainless steel and nickel alloy core shroud, core plate, core plate bolts, support structure, top guide, core spray lines, spargers, jet pump assemblies, control rod drive housing, nuclear instrumentation guide tubes	Cracking due to stress corrosion cracking, intergranular stress corrosion cracking, irradiation-assisted stress corrosion cracking	BWR Vessel Internals and Water Chemistry	No	R-92 R-93 R-96 R-97 R-98 R-99 R-100 R-105	IV.B1-1 IV.B1-6 IV.B1-2 IV.B1-3 IV.B1-17 IV.B1-7 IV.B1-13 IV.B1-10
45	BWR	Steel piping, piping components, and piping elements exposed to reactor coolant	Wall thinning due to flow-accelerated corrosion	Flow-Accelerated Corrosion	No	R-23	IV.C1-7
46	BWR	Nickel alloy core shroud and core plate access hole cover (mechanical covers)	Cracking due to stress corrosion cracking, intergranular stress corrosion cracking, irradiation-assisted stress corrosion cracking	Inservice Inspection (IWB, IWC, and IWD), and Water Chemistry	No	R-95	IV.B1-4
47	BWR	Stainless steel and nickel-alloy reactor vessel internals exposed to reactor coolant	Loss of material due to pitting and crevice corrosion	Inservice Inspection (IWB, IWC, and IWD), and Water Chemistry	No	RP-26	IV.B1-15

Table 1. Summary of Aging Management Programs for the Reactor Coolant System Evaluated in Chapter IV of the GALL Report

ID	Type	Component	Aging Effect/Mechanism	Aging Management Programs	Further Evaluation Recommended	Related Generic Item	Unique Item
48	BWR	Steel and stainless steel Class 1 piping, fittings and branch connections < NPS 4 exposed to reactor coolant	Cracking due to stress corrosion cracking, intergranular stress corrosion cracking (for stainless steel only), and thermal and mechanical loading	Inservice Inspection (IWB, IWC, and IWD), Water chemistry, and One-Time Inspection of ASME Code Class 1 Small-bore Piping	No	R-03	IV.C1-1
49	BWR	Nickel alloy core shroud and core plate access hole cover (welded covers)	Cracking due to stress corrosion cracking, intergranular stress corrosion cracking, irradiation-assisted stress corrosion cracking	Inservice Inspection (IWB, IWC, and IWD), Water Chemistry, and, for BWRs with a crevice in the access hole covers, augmented inspection using UT or other demonstrated acceptable inspection of the access hole cover welds	No	R-94	IV.B1-5
50	BWR	High-strength low alloy steel top head closure studs and nuts exposed to air with reactor coolant leakage	Cracking due to stress corrosion cracking and intergranular stress corrosion cracking	Reactor Head Closure Studs	No	R-60	IV.A1-9

Table 1. Summary of Aging Management Programs for the Reactor Coolant System Evaluated in Chapter IV of the GALL Report

ID	Type	Component	Aging Effect/Mechanism	Aging Management Programs	Further Evaluation Recommended	Related Generic Item	Unique Item
51	BWR	Cast austenitic stainless steel jet pump assembly castings; orificed fuel support	Loss of fracture toughness due to thermal aging and neutron irradiation embrittlement	Thermal Aging and Neutron Irradiation Embrittlement of CASS	No	R-101 R-103	IV.B1-11 IV.B1-9
52	BWR/ PWR	Steel and stainless steel reactor coolant pressure boundary (RCPB) pump and valve closure bolting, manway and holding bolting, flange bolting, and closure bolting in high-pressure and high-temperature systems	Cracking due to stress corrosion cracking, loss of material due to wear, loss of preload due to thermal effects, gasket creep, and self-loosening	Bolting Integrity	No	R-10 R-11 R-12 R-26 R-27 R-29 R-32 R-78 R-79 R-80	IV.D1-2 IV.C2-7 IV.C2-8 IV.C1-12 IV.C1-10 IV.C1-13 IV.D1-10 IV.D2-6 IV.A2-6 IV.A2-7 IV.A2-8
53	BWR/ PWR	Steel piping, piping components, and piping elements exposed to closed cycle cooling water	Loss of material due to general, pitting and crevice corrosion	Closed-Cycle Cooling Water System	No	RP-10	IV.C2-14
54	BWR/ PWR	Copper alloy piping, piping components, and piping elements exposed to closed cycle cooling water	Loss of material due to pitting, crevice, and galvanic corrosion	Closed-Cycle Cooling Water System	No	RP-11	IV.C2-11

Table 1. Summary of Aging Management Programs for the Reactor Coolant System Evaluated in Chapter IV of the GALL Report

ID	Type	Component	Aging Effect/Mechanism	Aging Management Programs	Further Evaluation Recommended	Related Generic Item	Unique Item
55	BWR/ PWR	Cast austenitic stainless steel Class 1 pump casings, and valve bodies and bonnets exposed to reactor coolant >250°C (>482°F)	Loss of fracture toughness due to thermal aging embrittlement	Inservice inspection (IWB, IWC, and IWD). Thermal aging susceptibility screening is not necessary, inservice inspection requirements are sufficient for managing these aging effects. ASME Code Case N-481 also provides an alternative for pump casings.	No	R-08	IV.C1-3 IV.C2-6
56	BWR/ PWR	Copper alloy >15% Zn piping, piping components, and piping elements exposed to closed cycle cooling water	Loss of material due to selective leaching	Selective Leaching of Materials	No	RP-12	IV.C2-12
57	BWR/ PWR	Cast austenitic stainless steel Class 1 piping, piping component, and piping elements and control rod drive pressure housings exposed to reactor coolant >250°C (>482°F)	Loss of fracture toughness due to thermal aging embrittlement	Thermal Aging Embrittlement of CASS	No	R-52 R-77	IV.C1-2 IV.C2-4 IV.A2-10
58	PWR	Steel reactor coolant pressure boundary external surfaces exposed to air with borated water leakage	Loss of material due to Boric acid corrosion	Boric Acid Corrosion	No	R-17	IV.A2-13 IV.C2-9 IV.D1-3 IV.D2-1

Table 1. Summary of Aging Management Programs for the Reactor Coolant System Evaluated in Chapter IV of the GALL Report

ID	Type	Component	Aging Effect/Mechanism	Aging Management Programs	Further Evaluation Recommended	Related Generic Item	Unique Item
59	PWR	Steel steam generator steam nozzle and safe end, feedwater nozzle and safe end, AFW nozzles and safe ends exposed to secondary feedwater/steam	Wall thinning due to flow-accelerated corrosion	Flow-Accelerated Corrosion	No	R-37 R-38	IV.D1-5 IV.D2-7
60	PWR	Stainless steel flux thimble tubes (with or without chrome plating)	Loss of material due to Wear	Flux Thimble Tube Inspection	No	R-145	IV.B2-13
61	PWR	Stainless steel, steel pressurizer integral support exposed to air with metal temperature up to 288°C (550°F)	Cracking due to cyclic loading	Inservice Inspection (IWB, IWC, and IWD)	No	R-19	IV.C2-16
62	PWR	Stainless steel, steel with stainless steel cladding reactor coolant system cold leg, hot leg, surge line, and spray line piping and fittings exposed to reactor coolant	Cracking due to cyclic loading	Inservice Inspection (IWB, IWC, and IWD)	No	R-56	IV.C2-26
63	PWR	Steel reactor vessel flange, stainless steel and nickel alloy reactor vessel internals exposed to reactor coolant (e.g., upper and lower internals assembly, CEA shroud assembly, core support barrel, upper grid assembly, core support shield assembly, lower grid assembly)	Loss of material due to Wear	Inservice Inspection (IWB, IWC, and IWD)	No	R-87 R-115 R-142 R-148 R-152 R-156 R-170 R-179 R-190 R-208	IV.A2-25 IV.B2-34 IV.B2-26 IV.B3-26 IV.B3-3 IV.B3-17 IV.B3-22 IV.B4-42 IV.B4-15 IV.B4-27

Table 1. Summary of Aging Management Programs for the Reactor Coolant System Evaluated in Chapter IV of the GALL Report

ID	Type	Component	Aging Effect/Mechanism	Aging Management Programs	Further Evaluation Recommended	Related Generic Item	Unique Item
64	PWR	Stainless steel and steel with stainless steel or nickel alloy cladding pressurizer components	Cracking due to stress corrosion cracking, primary water stress corrosion cracking	Inservice Inspection (IWB, IWC, and IWD) and Water Chemistry	No	R-25	IV.C2-19
65	PWR	Nickel alloy reactor vessel upper head and control rod drive penetration nozzles, instrument tubes, head vent pipe (top head), and welds	Cracking due to primary water stress corrosion cracking	Inservice Inspection (IWB, IWC, and IWD) and Water Chemistry and Nickel-Alloy Penetration Nozzles Welded to the Upper Reactor Vessel Closure Heads of Pressurized Water Reactors	No	R-75 R-90	IV.A2-9 IV.A2-18
66	PWR	Steel steam generator secondary manways and handholds (cover only) exposed to air with leaking secondary-side water and/or steam	Loss of material due to erosion	Inservice Inspection (IWB, IWC, and IWD) for Class 2 components	No	R-31	IV.D2-5
67	PWR	Steel with stainless steel or nickel alloy cladding; or stainless steel pressurizer components exposed to reactor coolant	Cracking due to cyclic loading	Inservice Inspection (IWB, IWC, and IWD), and Water Chemistry	No	R-58	IV.C2-18

Table 1. Summary of Aging Management Programs for the Reactor Coolant System Evaluated in Chapter IV of the GALL Report

ID	Type	Component	Aging Effect/Mechanism	Aging Management Programs	Further Evaluation Recommended	Related Generic Item	Unique Item
68	PWR	Stainless steel, steel with stainless steel cladding Class 1 piping, fittings, pump casings, valve bodies, nozzles, safe ends, manways, flanges, CRD housing; pressurizer heater sheaths, sleeves, diaphragm plate; pressurizer relief tank components, reactor coolant system cold leg, hot leg, surge line, and spray line piping and fittings	Cracking due to stress corrosion cracking	Inservice Inspection (IWB, IWC, and IWD), and Water Chemistry	No	R-07 R-09 R-14 R-30 R-217	IV.C2-2 IV.D1-1 IV.C2-5 IV.C2-22 IV.C2-27 IV.C2-20
69	PWR	Stainless steel, nickel alloy safety injection nozzles, safe ends, and associated welds and buttering exposed to reactor coolant	Cracking due to stress corrosion cracking, primary water stress corrosion cracking	Inservice Inspection (IWB, IWC, and IWD), and Water Chemistry	No	R-83	IV.A2-15
70	PWR	Stainless steel; steel with stainless steel cladding Class 1 piping, fittings and branch connections < NPS 4 exposed to reactor coolant	Cracking due to stress corrosion cracking, thermal and mechanical loading	Inservice Inspection (IWB, IWC, and IWD), Water chemistry, and One-Time Inspection of ASME Code Class 1 Small-bore Piping	No	R-02	IV.C2-1
71	PWR	High-strength low alloy steel closure head stud assembly exposed to air with reactor coolant leakage	Cracking due to stress corrosion cracking; loss of material due to wear	Reactor Head Closure Studs	No	R-71 R-72	IV.A2-2 IV.A2-3

Table 1. Summary of Aging Management Programs for the Reactor Coolant System Evaluated in Chapter IV of the GALL Report

ID	Type	Component	Aging Effect/Mechanism	Aging Management Programs	Further Evaluation Recommended	Related Generic Item	Unique Item
72	PWR	Nickel alloy steam generator tubes and sleeves exposed to secondary feedwater/ steam	Cracking due to OD stress corrosion cracking and intergranular attack, loss of material due to fretting and wear	Steam Generator Tube Integrity and Water Chemistry	No	R-47 R-48 R-49	IV.D1-23 IV.D2-17 IV.D1-22 IV.D2-16 IV.D1-24 IV.D2-18
73	PWR	Nickel alloy steam generator tubes, repair sleeves, and tube plugs exposed to reactor coolant	Cracking due to primary water stress corrosion cracking	Steam Generator Tube Integrity and Water Chemistry	No	R-40 R-44	IV.D1-18 IV.D2-12 IV.D1-20 IV.D2-14
74	PWR	Chrome plated steel, stainless steel, nickel alloy steam generator anti-vibration bars exposed to secondary feedwater/ steam	Cracking due to stress corrosion cracking, loss of material due to crevice corrosion and fretting	Steam Generator Tube Integrity and Water Chemistry	No	RP-14 RP-15	IV.D1-14 IV.D1-15
75	PWR	Nickel alloy once-through steam generator tubes exposed to secondary feedwater/ steam	Denting due to corrosion of carbon steel tube support plate	Steam Generator Tube Integrity and Water Chemistry	No	R-226	IV.D2-13
76	PWR	Steel steam generator tube support plate, tube bundle wrapper exposed to secondary feedwater/steam	Loss of material due to erosion, general, pitting, and crevice corrosion, ligament cracking due to corrosion	Steam Generator Tube Integrity and Water Chemistry	No	R-42 RP-16	IV.D1-17 IV.D2-11 IV.D1-9

Table 1. Summary of Aging Management Programs for the Reactor Coolant System Evaluated in Chapter IV of the GALL Report

ID	Type	Component	Aging Effect/Mechanism	Aging Management Programs	Further Evaluation Recommended	Related Generic Item	Unique Item
77	PWR	Nickel alloy steam generator tubes and sleeves exposed to phosphate chemistry in secondary feedwater/ steam	Loss of material due to wastage and pitting corrosion	Steam Generator Tube Integrity and Water Chemistry	No	R-50	IV.D1-25
78	PWR	Steel steam generator tube support lattice bars exposed to secondary feedwater/ steam	Wall thinning due to flow-accelerated corrosion	Steam Generator Tube Integrity and Water Chemistry	No	R-41	IV.D1-16
79	PWR	Nickel alloy steam generator tubes exposed to secondary feedwater/ steam	Denting due to corrosion of steel tube support plate	Steam Generator Tube Integrity; Water Chemistry and, for plants that could experience denting at the upper support plates, evaluate potential for rapidly propagating cracks and then develop and take corrective actions consistent with Bulletin 88-02.	No	R-43	IV.D1-19
80	PWR	Cast austenitic stainless steel reactor vessel internals (e.g., upper internals assembly, lower internal assembly, CEA shroud assemblies, control rod guide tube assembly, core support shield assembly, lower grid assembly)	Loss of fracture toughness due to thermal aging and neutron irradiation embrittlement	Thermal Aging and Neutron Irradiation Embrittlement of CASS	No	R-111 R-140 R-153 R-171 R-183 R-191 R-206	IV.B2-37 IV.B2-21 IV.B3-1 IV.B3-18 IV.B44 IV.B4-21 IV.B4-28

Table 1. Summary of Aging Management Programs for the Reactor Coolant System Evaluated in Chapter IV of the GALL Report

ID	Type	Component	Aging Effect/Mechanism	Aging Management Programs	Further Evaluation Recommended	Related Generic Item	Unique Item
81	PWR	Nickel alloy or nickel-alloy clad steam generator divider plate exposed to reactor coolant	Cracking due to primary water stress corrosion cracking	Water Chemistry	No	RP-21	IV.D1-6
82	PWR	Stainless steel steam generator primary side divider plate exposed to reactor coolant	Cracking due to stress corrosion cracking	Water Chemistry	No	RP-17	IV.D1-7
83	PWR	Stainless steel; steel with nickel-alloy or stainless steel cladding; and nickel-alloy reactor vessel internals and reactor coolant pressure boundary components exposed to reactor coolant	Loss of material due to pitting and crevice corrosion	Water Chemistry	No	RP-23 RP-24 RP-28	IV.C2-15 IV.B2-32 IV.B3-25 IV.B4-38 IV.A2-14
84	PWR	Nickel alloy steam generator components such as, secondary side nozzles (vent, drain, and instrumentation) exposed to secondary feedwater/ steam	Cracking due to stress corrosion cracking	Water Chemistry and One-Time Inspection or Inservice Inspection (IWB, IWC, and IWD).	No	R-36	IV.D2-9
85	BWR/ PWR	Nickel alloy piping, piping components, and piping elements exposed to air – indoor uncontrolled (external)	None	None	NA - No AEM or AMP	RP-03	IV.E-1
86	BWR/ PWR	Stainless steel piping, piping components, and piping elements exposed to air – indoor uncontrolled (External); air with borated water leakage; concrete; gas	None	None	NA - No AEM or AMP	RP-04 RP-05 RP-06 RP-07	IV.E-2 IV.E-3 IV.E-4 IV.E-5

Table 1. Summary of Aging Management Programs for the Reactor Coolant System Evaluated in Chapter IV of the GALL Report

ID	Type	Component	Aging Effect/Mechanism	Aging Management Programs	Further Evaluation Recommended	Related Generic Item	Unique Item
87	BWR/ PWR	Steel piping, piping components, and piping elements in concrete	None	None	NA - No AEM or AMP	RP-01	IV.E-6

Table 2. Summary of Aging Management Programs for the Engineered Safety Features Evaluated in Chapter V of the GALL Report

ID	Type	Component	Aging Effect/Mechanism	Aging Management Programs	Further Evaluation Recommended	Related Generic Item	Unique Item
1	BWR/ PWR	Steel and stainless steel piping, piping components, and piping elements in emergency core cooling system	Cumulative fatigue damage	TLAA, evaluated in accordance with 10 CFR 54.21(c)	Yes, TLAA	E-10 E-13	V.D2-32 V.D1-27
2	PWR	Steel with stainless steel cladding pump casing exposed to treated borated water	Loss of material/ cladding breach	A plant-specific aging management program is to be evaluated. Reference NRC Information Notice 94-63, "Boric Acid Corrosion of Charging Pump Casings Caused by Cladding Cracks."	Yes, verify that plant-specific program addresses cladding breach	EP-49	V.D1-32
3	BWR/ PWR	Stainless steel containment isolation piping and components internal surfaces exposed to treated water	Loss of material due to pitting and crevice corrosion	Water Chemistry and One-Time Inspection	Yes, detection of aging effects is to be evaluated	E-33	V.C-4
4	BWR/ PWR	Stainless steel piping, piping components, and piping elements exposed to soil	Loss of material due to pitting and crevice corrosion	A plant-specific aging management program is to be evaluated.	Yes, plant specific	EP-31	V.D1-26 V.D2-27

Table 2.		Summary of Aging Management Programs for the Engineered Safety Features Evaluated in Chapter V of the GALL Report				
ID	Type	Component	Aging Effect/Mechanism	Aging Management Programs	Further Evaluation Recommended	Related Generic Item	Unique Item
5	BWR	Stainless steel and aluminum piping, piping components, and piping elements exposed to treated water	Loss of material due to pitting and crevice corrosion	Water Chemistry and One-Time Inspection	Yes, detection of aging effects is to be evaluated	EP-26 EP-32	V.D2-19 V.D2-28
6	BWR/ PWR	Stainless steel and copper alloy piping, piping components, and piping elements exposed to lubricating oil	Loss of material due to pitting and crevice corrosion	Lubricating Oil Analysis and One-Time Inspection	Yes, detection of aging effects is to be evaluated	EP-45 EP-51	V.A-21 V.D1-18 V.D2-22 V.D1-24
7	BWR/ PWR	Partially encased stainless steel tanks with breached moisture barrier exposed to raw water	Loss of material due to pitting and crevice corrosion	A plant-specific aging management program is to be evaluated for pitting and crevice corrosion of tank bottoms because moisture and water can egress under the tank due to cracking of the perimeter seal from weathering.	Yes, plant specific	E-01	V.D1-15

Table 2. Summary of Aging Management Programs for the Engineered Safety Features Evaluated in Chapter V of the GALL Report

ID	Type	Component	Aging Effect/Mechanism	Aging Management Programs	Further Evaluation Recommended	Related Generic Item	Unique Item
8	BWR/ PWR	Stainless steel piping, piping components, piping elements, and tank internal surfaces exposed to condensation (internal)	Loss of material due to pitting and crevice corrosion	A plant-specific aging management program is to be evaluated.	Yes, plant specific	E-14 EP-53	V.D2-35 V.A-26 V.D1-29
9	BWR/ PWR	Steel, stainless steel, and copper alloy heat exchanger tubes exposed to lubricating oil	Reduction of heat transfer due to fouling	Lubricating Oil Analysis and One-Time Inspection	Yes, detection of aging effects is to be evaluated	EP-40 EP-47 EP-50	V.A-17 V.D1-12 V.D2-14 V.A-12 V.D1-8 V.D2-9 V.A-14 V.D1-10 V.D2-11
10	BWR/ PWR	Stainless steel heat exchanger tubes exposed to treated water	Reduction of heat transfer due to fouling	Water Chemistry and One-Time Inspection	Yes, detection of aging effects is to be evaluated	EP-34	V.A-16 V.D2-13
11	BWR	Elastomer seals and components in standby gas treatment system exposed to air - indoor uncontrolled	Hardening and loss of strength due to elastomer degradation	A plant-specific aging management program is to be evaluated.	Yes, plant specific 5)	E-06	V.B-4

Table 2. Summary of Aging Management Programs for the Engineered Safety Features Evaluated in Chapter V of the GALL Report

ID	Type	Component	Aging Effect/Mechanism	Aging Management Programs	Further Evaluation Recommended	Related Generic Item	Unique Item
12	PWR	Stainless steel high-pressure safety injection (charging) pump miniflow orifice exposed to treated borated water	Loss of material due to erosion	A plant-specific aging management program is to be evaluated for erosion of the orifice due to extended use of the centrifugal HPSI pump for normal charging.	Yes, plant specific	E-24	V.D1-14
13	BWR	Steel drywell and suppression chamber spray system nozzle and flow orifice internal surfaces exposed to air – indoor uncontrolled (internal)	Loss of material due to general corrosion and fouling	A plant-specific aging management program is to be evaluated.	Yes, plant specific	E-04	V.D2-1
14	BWR	Steel piping, piping components, and piping elements exposed to treated water	Loss of material due to general, pitting, and crevice corrosion	Water Chemistry and One-Time Inspection	Yes, detection of aging effects is to be evaluated	E-08	V.D2-33
15	BWR/ PWR	Steel containment isolation piping, piping components, and piping elements internal surfaces exposed to treated water	Loss of material due to general, pitting, and crevice corrosion	Water Chemistry and One-Time Inspection	Yes, detection of aging effects is to be evaluated	E-31	V.C-6

Table 2. Summary of Aging Management Programs for the Engineered Safety Features Evaluated in Chapter V of the GALL Report

ID	Type	Component	Aging Effect/Mechanism	Aging Management Programs	Further Evaluation Recommended	Related Generic Item	Unique Item
16	BWR/ PWR	Steel piping, piping components, and piping elements exposed to lubricating oil	Loss of material due to general, pitting, and crevice corrosion	Lubricating Oil Analysis and One-Time Inspection	Yes, detection of aging effects is to be evaluated	EP-46	V.A-25 V.D1-28 V.D2-30
17	BWR/ PWR	Steel (with or without coating or wrapping) piping, piping components, and piping elements buried in soil	Loss of material due to general, pitting, crevice, and microbiologically-influenced corrosion	Buried Piping and Tanks Surveillance or Buried Piping and Tanks Inspection	No Yes, detection of aging effects and operating experience are to be further evaluated	E-42	V.B-9
18	BWR	Stainless steel piping, piping components, and piping elements exposed to treated water >60°C (>140°F)	Cracking due to stress corrosion cracking and intergranular stress corrosion cracking	BWR Stress Corrosion Cracking and Water Chemistry	No	E-37	V.D2-29
19	BWR	Steel piping, piping components, and piping elements exposed to steam or treated water	Wall thinning due to flow-accelerated corrosion	Flow-Accelerated Corrosion	No	E-07 E-09	V.D2-31 V.D2-34

Table 2. Summary of Aging Management Programs for the Engineered Safety Features Evaluated in Chapter V of the GALL Report

ID	Type	Component	Aging Effect/Mechanism	Aging Management Programs	Further Evaluation Recommended	Related Generic Item	Unique Item
20	BWR	Cast austenitic stainless steel piping, piping components, and piping elements exposed to treated water (borated or unborated) >250°C (>482°F)	Loss of fracture toughness due to thermal aging embrittlement	Thermal Aging Embrittlement of CASS	No	E-11	V.D2-20
21	BWR/PWR	High-strength steel closure bolting exposed to air with steam or water leakage	Cracking due to cyclic loading, stress corrosion cracking	Bolting Integrity	No	E-03	V.E-3
22	BWR/PWR	Steel closure bolting exposed to air with steam or water leakage	Loss of material due to general corrosion	Bolting Integrity	No	E-02	V.E-6
23	BWR/PWR	Steel bolting and closure bolting exposed to air – outdoor (external), or air – indoor uncontrolled (external)	Loss of material due to general, pitting, and crevice corrosion	Bolting Integrity	No	EP-1 EP-25	V.E-1 V.E-4
24	BWR/PWR	Steel closure bolting exposed to air – indoor uncontrolled (external)	Loss of preload due to thermal effects, gasket creep, and self-loosening	Bolting Integrity	No	EP-24	V.E-5

Table 2. Summary of Aging Management Programs for the Engineered Safety Features Evaluated in Chapter V of the GALL Report

ID	Type	Component	Aging Effect/Mechanism	Aging Management Programs	Further Evaluation Recommended	Related Generic Item	Unique Item
25	BWR/ PWR	Stainless steel piping, piping components, and piping elements exposed to closed cycle cooling water >60°C (>140°F)	Cracking due to stress corrosion cracking	Closed-Cycle Cooling Water System	No	EP-44	V.A-24 V.C-8 V.D1-23 V.D2-26
26	BWR/ PWR	Steel piping, piping components, and piping elements exposed to closed cycle cooling water	Loss of material due to general, pitting, and crevice corrosion	Closed-Cycle Cooling Water System	No	EP-48	V.C-9
27	BWR/ PWR	Steel heat exchanger components exposed to closed cycle cooling water	Loss of material due to general, pitting, crevice, and galvanic corrosion	Closed-Cycle Cooling Water System	No	E-17	V.A-9 V.D1-6 V.D2-7
28	BWR/ PWR	Stainless steel piping, piping components, piping elements, and heat exchanger components exposed to closed-cycle cooling water	Loss of material due to pitting and crevice corrosion	Closed-Cycle Cooling Water System	No	E-19 EP-33	V.A-7 V.D1-4 V.D2-5 V.A-23 V.C-7 V.D1-22 V.D2-25

Table 2. Summary of Aging Management Programs for the Engineered Safety Features Evaluated in Chapter V of the GALL Report

ID	Type	Component	Aging Effect/Mechanism	Aging Management Programs	Further Evaluation Recommended	Related Generic Item	Unique Item
29	BWR/ PWR	Copper alloy piping, piping components, piping elements, and heat exchanger components exposed to closed cycle cooling water	Loss of material due to pitting, crevice, and galvanic corrosion	Closed-Cycle Cooling Water System	No	EP-13 EP-36	V.A-5 V.D1-2 V.D2-3 V.A-20 V.B-6 V.D1-17 V.D2-21
30	BWR/ PWR	Stainless steel and copper alloy heat exchanger tubes exposed to closed cycle cooling water	Reduction of heat transfer due to fouling	Closed-Cycle Cooling Water System	No	EP-35 EP-39	V.A-13 V.D1-9 V.D2-10 V.A-11
31	BWR/ PWR	External surfaces of steel components including ducting, piping, ducting closure bolting, and containment isolation piping external surfaces exposed to air – indoor uncontrolled (external); condensation (external) and air - outdoor (external)	Loss of material due to general corrosion	External Surfaces Monitoring	No	E-26 E-30 E-35 E-40 E-44 E-45 E-46	V.A-1 V.B-3 V.D2-2 V.C-2 V.C-1 V.B-2 V.E-7 V.E-8 V.E-10

Table 2. Summary of Aging Management Programs for the Engineered Safety Features Evaluated in Chapter V of the GALL Report

ID	Type	Component	Aging Effect/Mechanism	Aging Management Programs	Further Evaluation Recommended	Related Generic Item	Unique Item
32	BWR/ PWR	Steel piping and ducting components and internal surfaces exposed to air – indoor uncontrolled (Internal)	Loss of material due to general corrosion	Inspection of Internal Surfaces in Miscellaneous Piping and Ducting Components	No	E-25 E-29	V.B-1 V.A-19 V.D2-16
33	BWR/ PWR	Steel encapsulation components exposed to air – indoor uncontrolled (internal)	Loss of material due to general, pitting, and crevice corrosion	Inspection of Internal Surfaces in Miscellaneous Piping and Ducting Components	No	EP-42	V.A-2
34	BWR/ PWR	Steel piping, piping components, and piping elements exposed to condensation (internal)	Loss of material due to general, pitting, and crevice corrosion	Inspection of Internal Surfaces in Miscellaneous Piping and Ducting Components	No	E-27	V.D2-17
35	BWR/ PWR	Steel containment isolation piping and components internal surfaces exposed to raw water	Loss of material due to general, pitting, crevice, and microbiologically-influenced corrosion, and fouling	Open-Cycle Cooling Water System	No	E-22	V.C-5

Table 2. Summary of Aging Management Programs for the Engineered Safety Features Evaluated in Chapter V of the GALL Report

ID	Type	Component	Aging Effect/Mechanism	Aging Management Programs	Further Evaluation Recommended	Related Generic Item	Unique Item
36	BWR/ PWR	Steel heat exchanger components exposed to raw water	Loss of material due to general, pitting, crevice, galvanic, and microbiologically-influenced corrosion, and fouling	Open-Cycle Cooling Water System	No	E-18	V.A-10 V.D1-7 V.D2-8
37	BWR/ PWR	Stainless steel piping, piping components, and piping elements exposed to raw water	Loss of material due to pitting, crevice, and microbiologically-influenced corrosion	Open-Cycle Cooling Water System	No	EP-55	V.D1-25
38	BWR/ PWR	Stainless steel containment isolation piping and components internal surfaces exposed to raw water	Loss of material due to pitting, crevice, and microbiologically-influenced corrosion, and fouling	Open-Cycle Cooling Water System	No	E-34	V.C-3
39	BWR/ PWR	Stainless steel heat exchanger components exposed to raw water	Loss of material due to pitting, crevice, and microbiologically-influenced corrosion, and fouling	Open-Cycle Cooling Water System	No	E-20	V.A-8 V.D1-5 V.D2-6

Table 2. Summary of Aging Management Programs for the Engineered Safety Features Evaluated in Chapter V of the GALL Report

ID	Type	Component	Aging Effect/Mechanism	Aging Management Programs	Further Evaluation Recommended	Related Generic Item	Unique Item
40	BWR/ PWR	Steel and stainless steel heat exchanger tubes (serviced by open-cycle cooling water) exposed to raw water	Reduction of heat transfer due to fouling	Open-Cycle Cooling Water System	No	E-21 E-23	V.A-15 V.D1-11 V.D2-12 V.D2-15
41	BWR/ PWR	Copper alloy >15% Zn piping, piping components, piping elements, and heat exchanger components exposed to closed cycle cooling water	Loss of material due to selective leaching	Selective Leaching of Materials	No	EP-27 EP-37	V.A-22 V.B-7 V.D1-19 V.D2-23 V.A-6 V.B-5 V.D1-3 V.D2-4
42	BWR/ PWR	Gray cast iron piping, piping components, piping elements exposed to closed-cycle cooling water	Loss of material due to selective leaching	Selective Leaching of Materials	No	EP-52	V.D1-20
43	BWR/ PWR	Gray cast iron piping, piping components, and piping elements exposed to soil	Loss of material due to selective leaching	Selective Leaching of Materials	No	EP-54	V.B-8 V.D1-21 V.D2-24
44	BWR/ PWR	Gray cast iron motor cooler exposed to treated water	Loss of material due to selective leaching	Selective Leaching of Materials	No	E-43	V.A-18 V.D1-13

Table 2. Summary of Aging Management Programs for the Engineered Safety Features Evaluated in Chapter V of the GALL Report

ID	Type	Component	Aging Effect/Mechanism	Aging Management Programs	Further Evaluation Recommended	Related Generic Item	Unique Item
45	PWR	Aluminum, copper alloy >15% Zn, and steel external surfaces, bolting, and piping, piping components, and piping elements exposed to air with borated water leakage	Loss of material due to Boric acid corrosion	Boric Acid Corrosion	No	E-28 E-41 EP-2 EP-38	V.A-4 V.D1-1 V.E-9 V.E-2 V.D2-18 V.E-11
46	PWR	Steel encapsulation components exposed to air with borated water leakage (internal)	Loss of material due to general, pitting, crevice and boric acid corrosion	Inspection of Internal Surfaces in Miscellaneous Piping and Ducting Components	No	EP-43	V.A-3
47	PWR	Cast austenitic stainless steel piping, piping components, and piping elements exposed to treated borated water >250°C (>482°F)	Loss of fracture toughness due to thermal aging embrittlement	Thermal Aging Embrittlement of CASS	No	E-47	V.D1-16

Table 2. Summary of Aging Management Programs for the Engineered Safety Features Evaluated in Chapter V of the GALL Report

ID	Type	Component	Aging Effect/Mechanism	Aging Management Programs	Further Evaluation Recommended	Related Generic Item	Unique Item
48	PWR	Stainless steel or stainless-steel-clad steel piping, piping components, piping elements, and tanks (including safety injection tanks/accumulators) exposed to treated borated water >60°C (>140°F)	Cracking due to stress corrosion cracking	Water Chemistry	No	E-12 E-38	V.A-28 V.D1-31 V.D1-33
49	PWR	Stainless steel piping, piping components, piping elements, and tanks exposed to treated borated water	Loss of material due to pitting and crevice corrosion	Water Chemistry	No	EP-41	V.A-27 V.D1-30
50	BWR/ PWR	Aluminum piping, piping components, and piping elements exposed to air-indoor uncontrolled (internal/external)	None	None	NA - No AEM or AMP	EP-3	V.F-2
51	BWR/ PWR	Galvanized steel ducting exposed to air – indoor controlled (external)	None	None	NA - No AEM or AMP	EP-14	V.F-1

Table 2. Summary of Aging Management Programs for the Engineered Safety Features Evaluated in Chapter V of the GALL Report

ID	Type	Component	Aging Effect/Mechanism	Aging Management Programs	Further Evaluation Recommended	Related Generic Item	Unique Item
52	BWR/PWR	Glass piping elements exposed to air – indoor uncontrolled (external), lubricating oil, raw water, treated water, or treated borated water	None	None	NA - No AEM or AMP	EP-15 EP-16 EP-28 EP-29 EP-30	V.F-6 V.F-7 V.F-8 V.F-10 V.F-9
53	BWR/PWR	Stainless steel, copper alloy, and nickel alloy piping, piping components, and piping elements exposed to air – indoor uncontrolled (external)	None	None	NA - No AEM or AMP	EP-10 EP-17 EP-18	V.F-3 V.F-11 V.F-12
54	BWR/PWR	Steel piping, piping components, and piping elements exposed to air – indoor controlled (external)	None	None	NA - No AEM or AMP	EP-4	V.F-16
55	BWR/PWR	Steel and stainless steel piping, piping components, and piping elements in concrete	None	None	NA - No AEM or AMP	EP-5 EP-20	V.F-17 V.F-14

Table 2. Summary of Aging Management Programs for the Engineered Safety Features Evaluated in Chapter V of the GALL Report

ID	Type	Component	Aging Effect/Mechanism	Aging Management Programs	Further Evaluation Recommended	Related Generic Item	Unique Item
56	BWR/ PWR	Steel, stainless steel, and copper alloy piping, piping components, and piping elements exposed to gas	None	None	NA - No AEM or AMP	EP-7 EP-9 EP-22	V.F-18 V.F-4 V.F-15
57	PWR	Stainless steel and copper alloy <15% Zn piping, piping components, and piping elements exposed to air with borated water leakage	None	None	NA - No AEM or AMP	EP-12 EP-19	V.F-5 V.F-13

Table 3. Summary of Aging Management Programs for the Auxiliary Systems Evaluated in Chapter VII of the GALL Report

ID	Type	Component	Aging Effect/Mechanism	Aging Management Programs	Further Evaluation Recommended	Related Generic Item	Unique Item
1	BWR/ PWR	Steel cranes - structural girders exposed to air – indoor uncontrolled (external)	Cumulative fatigue damage	TLAA to be evaluated for structural girders of cranes. See the Standard Review Plan, Section 4.7 for generic guidance for meeting the requirements of 10 CFR 54.21(c)(1).	Yes, TLAA	A-06	VII.B-2
2	BWR/ PWR	Steel and stainless steel piping, piping components, piping elements, and heat exchanger components exposed to air – indoor uncontrolled, treated borated water or treated water	Cumulative fatigue damage	TLAA, evaluated in accordance with 10 CFR 54.21(c)	Yes, TLAA	A-34 A-57 A-62 A-100	VII.E1-18 VII.E3-17 VII.E1-16 VII.E3-14 VII.E4-13 VII.E1-4
3	BWR/ PWR	Stainless steel heat exchanger tubes exposed to treated water	Reduction of heat transfer due to fouling	Water Chemistry and One-Time Inspection	Yes, detection of aging effects is to be evaluated	AP-62	VII.A4-4 VII.E3-6
4	BWR	Stainless steel piping, piping components, and piping elements exposed to sodium pentaborate solution >60°C (>140°F)	Cracking due to stress corrosion cracking	Water Chemistry and One-Time Inspection	Yes, detection of aging effects is to be evaluated	A-59	VII.E2-2
5	BWR/ PWR	Stainless steel and stainless clad steel heat exchanger components exposed to treated water >60°C (>140°F)	Cracking due to stress corrosion cracking	Plant specific	Yes, plant specific	A-71 A-85	VII.E3-3 VII.E3-19

Table 3. Summary of Aging Management Programs for the Auxiliary Systems Evaluated in Chapter VII of the GALL Report

ID	Type	Component	Aging Effect/Mechanism	Aging Management Programs	Further Evaluation Recommended	Related Generic Item	Unique Item
6	BWR/ PWR	Stainless steel diesel engine exhaust piping, piping components, and piping elements exposed to diesel exhaust	Cracking due to stress corrosion cracking	Plant specific	Yes, plant specific	AP-33	VII.H2-1
7	PWR	Stainless steel non-regenerative heat exchanger components exposed to treated borated water >60°C (>140°F)	Cracking due to stress corrosion cracking and cyclic loading	Water Chemistry and a plant-specific verification program. An acceptable verification program is to include temperature and radioactivity monitoring of the shell side water, and eddy current testing of tubes.	Yes, plant specific	A-69	VII.E1-9
8	PWR	Stainless steel regenerative heat exchanger components exposed to treated borated water >60°C (>140°F)	Cracking due to stress corrosion cracking and cyclic loading	Water Chemistry and a plant-specific verification program. The AMP is to be augmented by verifying the absence of cracking due to stress corrosion cracking and cyclic loading. A plant specific aging management program is to be evaluated.	Yes, plant specific	A-84	VII.E1-5

Table 3. Summary of Aging Management Programs for the Auxiliary Systems Evaluated in Chapter VII of the GALL Report

ID	Type	Component	Aging Effect/Mechanism	Aging Management Programs	Further Evaluation Recommended	Related Generic Item	Unique Item
9	PWR	Stainless steel high-pressure pump casing in PWR chemical and volume control system	Cracking due to stress corrosion cracking and cyclic loading	Water Chemistry and a plant-specific verification program. The AMP is to be augmented by verifying the absence of cracking due to stress corrosion cracking and cyclic loading. A plant specific aging management program is to be evaluated.	Yes, plant specific	A-76	VII.E1-7
10	BWR/ PWR	High-strength steel closure bolting exposed to air with steam or water leakage.	Cracking due to stress corrosion cracking, cyclic loading	Bolting Integrity The AMP is to be augmented by appropriate inspection to detect cracking if the bolts are not otherwise replaced during maintenance.	Yes, if the bolts are not replaced during maintenance	A-104	VII.E1-8
11	BWR/ PWR	Elastomer seals and components exposed to air – indoor uncontrolled (internal/external)	Hardening and loss of strength due to elastomer degradation	Plant specific	Yes, plant specific	A-17	VII.F1-7 VII.F2-7 VII.F3-7 VII.F4-6
12	BWR/ PWR	Elastomer lining exposed to treated water or treated borated water	Hardening and loss of strength due to elastomer degradation	A plant-specific aging management program that determines and assesses the qualified life of the linings in the environment is to be evaluated.	Yes, plant specific	A-15 A-16	VII.A3-1 VII.A4-1

Table 3. Summary of Aging Management Programs for the Auxiliary Systems Evaluated in Chapter VII of the GALL Report

ID	Type	Component	Aging Effect/Mechanism	Aging Management Programs	Further Evaluation Recommended	Related Generic Item	Unique Item
13	BWR/ PWR	Boral, boron steel spent fuel storage racks neutron-absorbing sheets exposed to treated water or treated borated water	Reduction of neutron-absorbing capacity and loss of material due to general corrosion	Plant specific	Yes, plant specific	A-88 A-89	VII.A2-5 VII.A2-3
14	BWR/ PWR	Steel piping, piping component, and piping elements exposed to lubricating oil	Loss of material due to general, pitting, and crevice corrosion	Lubricating Oil Analysis and One-Time Inspection	Yes, detection of aging effects is to be evaluated	AP-30	VII.C1-17 VII.C2-13 VII.E1-19 VII.E4-16 VII.F1-19 VII.F2-17 VII.F3-19 VII.F4-15 VII.G-22 VII.H2-20
15	BWR/ PWR	Steel reactor coolant pump oil collection system piping, tubing, and valve bodies exposed to lubricating oil	Loss of material due to general, pitting, and crevice corrosion	Lubricating Oil Analysis and One-Time Inspection	Yes, detection of aging effects is to be evaluated	A-83	VII.G-26
16	BWR/ PWR	Steel reactor coolant pump oil collection system tank exposed to lubricating oil	Loss of material due to general, pitting, and crevice corrosion	Lubricating Oil Analysis and One-Time Inspection to evaluate the thickness of the lower portion of the tank	Yes, detection of aging effects is to be evaluated	A-82	VII.G-27
17	BWR	Steel piping, piping components, and piping elements exposed to treated water	Loss of material due to general, pitting, and crevice corrosion	Water Chemistry and One-Time Inspection	Yes, detection of aging effects is to be evaluated	A-35	VII.E3-18 VII.E4-17

Table 3. Summary of Aging Management Programs for the Auxiliary Systems Evaluated in Chapter VII of the GALL Report

ID	Type	Component	Aging Effect/Mechanism	Aging Management Programs	Further Evaluation Recommended	Related Generic Item	Unique Item
18	BWR/ PWR	Stainless steel and steel diesel engine exhaust piping, piping components, and piping elements exposed to diesel exhaust	Loss of material/ general (steel only), pitting and crevice corrosion	Plant specific	Yes, plant specific	A-27	VII.H2-2
19	BWR/ PWR	Steel (with or without coating or wrapping) piping, piping components, and piping elements exposed to soil	Loss of material due to general, pitting, crevice, and microbiologically influenced corrosion	Buried Piping and Tanks Surveillance or Buried Piping and Tanks Inspection	No Yes, detection of aging effects and operating experience are to be further evaluated	A-01	VII.C1-18 VII.C3-9 VII.G-25 VII.H1-9
20	BWR/ PWR	Steel piping, piping components, piping elements, and tanks exposed to fuel oil	Loss of material due to general, pitting, crevice, and microbiologically influenced corrosion, and fouling	Fuel Oil Chemistry and One-Time Inspection	Yes, detection of aging effects is to be evaluated	A-30	VII.H1-10 VII.H2-24
21	BWR/ PWR	Steel heat exchanger components exposed to lubricating oil	Loss of material due to general, pitting, crevice, and microbiologically influenced corrosion, and fouling	Lubricating Oil Analysis and One-Time Inspection	Yes, detection of aging effects is to be evaluated	AP-39	VII.H2-5

Table 3. Summary of Aging Management Programs for the Auxiliary Systems Evaluated in Chapter VII of the GALL Report

ID	Type	Component	Aging Effect/Mechanism	Aging Management Programs	Further Evaluation Recommended	Related Generic Item	Unique Item
22	BWR/PWR	Steel with elastomer lining or stainless steel cladding piping, piping components, and piping elements exposed to treated water and treated borated water	Loss of material due to pitting and crevice corrosion (only for steel after lining/cladding degradation)	Water Chemistry and One-Time Inspection	Yes, detection of aging effects is to be evaluated	A-39 A-40	VII.A3-9 VII.A4-12
23	BWR	Stainless steel and steel with stainless steel cladding heat exchanger components exposed to treated water	Loss of material due to pitting and crevice corrosion	Water Chemistry and One-Time Inspection	Yes, detection of aging effects is to be evaluated	A-70	VII.A4-2
24	BWR/PWR	Stainless steel and aluminum piping, piping components, and piping elements exposed to treated water	Loss of material due to pitting and crevice corrosion	Water Chemistry and One-Time Inspection	Yes, detection of aging effects is to be evaluated	A-58 AP-38	VII.A4-11 VII.E3-15 VII.E4-14 VII.A4-5 VII.E3-7 VII.E4-4
25	BWR/PWR	Copper alloy HVAC piping, piping components, piping elements exposed to condensation (external)	Loss of material due to pitting and crevice corrosion	A plant-specific aging management program is to be evaluated.	Yes, plant specific	A-46	VII.F1-16 VII.F2-14 VII.F3-16 VII.F4-12
26	BWR/PWR	Copper alloy piping, piping components, and piping elements exposed to lubricating oil	Loss of material due to pitting and crevice corrosion	Lubricating Oil Analysis and One-Time Inspection	Yes, detection of aging effects is to be evaluated	AP-47	VII.C1-8 VII.C2-5 VII.E1-12 VII.E4-6 VII.G-11 VII.H2-10

Table 3. Summary of Aging Management Programs for the Auxiliary Systems Evaluated in Chapter VII of the GALL Report

ID	Type	Component	Aging Effect/Mechanism	Aging Management Programs	Further Evaluation Recommended	Related Generic Item	Unique Item
27	BWR/PWR	Stainless steel HVAC ducting and aluminum HVAC piping, piping components and piping elements exposed to condensation	Loss of material due to pitting and crevice corrosion	A plant-specific aging management program is to be evaluated.	Yes, plant specific	A-09, AP-74	VII.F1-1 VII.F2-1 VII.F3-1 VII.F1-14 VII.F2-12 VII.F3-14 VII.F4-10
28	BWR/PWR	Copper alloy fire protection piping, piping components, and piping elements exposed to condensation (internal)	Loss of material due to pitting and crevice corrosion	A plant-specific aging management program is to be evaluated.	Yes, plant specific	AP-78	VII.G-9
29	BWR/PWR	Stainless steel piping, piping components, and piping elements exposed to soil	Loss of material due to pitting and crevice corrosion	A plant-specific aging management program is to be evaluated.	Yes, plant specific	AP-56	VII.C1-16 VII.C3-8 VII.G-20 VII.H1-7 VII.H2-19
30	BWR	Stainless steel piping, piping components, and piping elements exposed to sodium pentaborate solution	Loss of material due to pitting and crevice corrosion	Water Chemistry and One-Time Inspection	Yes, detection of aging effects is to be evaluated	AP-73	VII.E2-1
31	BWR	Copper alloy piping, piping components, and piping elements exposed to treated water	Loss of material due to pitting, crevice, and galvanic corrosion	Water Chemistry and One-Time Inspection	Yes, detection of aging effects is to be evaluated	AP-64	VII.A4-7 VII.E3-9 VII.E4-7
32	BWR/PWR	Stainless steel, aluminum and copper alloy piping, piping components, and piping elements exposed to fuel oil	Loss of material due to pitting, crevice, and microbiologically influenced corrosion	Fuel Oil Chemistry and One-Time Inspection	Yes, detection of aging effects is to be evaluated	AP-35, AP-44, AP-54	VII.H1-1 VII.H2-7 VII.G-10 VII.H1-3 VII.H2-9 VII.G-17 VII.H1-6 VII.H2-16

Table 3. Summary of Aging Management Programs for the Auxiliary Systems Evaluated in Chapter VII of the GALL Report

ID	Type	Component	Aging Effect/Mechanism	Aging Management Programs	Further Evaluation Recommended	Related Generic Item	Unique Item
33	BWR/ PWR	Stainless steel piping, piping components, and piping elements exposed to lubricating oil	Loss of material due to pitting, crevice, and microbiologically influenced corrosion	Lubricating Oil Analysis and One-Time Inspection	Yes, detection of aging effects is to be evaluated	AP-59	VII.C1-14 VII.C2-12 VII.E1-15 VII.E4-12 VII.G-18 VII.H2-17
34	BWR/ PWR	Elastomer seals and components exposed to air – indoor uncontrolled (internal or external)	Loss of material due to Wear	Plant specific	Yes, plant specific	A-18 A-73	VII.F1-6 VII.F2-6 VII.F3-6 VII.F4-5 VII.F1-5 VII.F2-5 VII.F3-5 VII.F4-4
35	PWR	Steel with stainless steel cladding pump casing exposed to treated borated water	Loss of material/ cladding breach	A plant-specific aging management program is to be evaluated. Reference NRC Information Notice 94-63, "Boric Acid Corrosion of Charging Pump Casings Caused by Cladding Cracks."	Yes, verify plant-specific program addresses cladding breach	AP-85	VII.E1-21
36	BWR	Boraflex spent fuel storage racks neutron-absorbing sheets exposed to treated water	Reduction of neutron-absorbing capacity due to boraflex degradation	Boraflex Monitoring	No	A-87	VII.A2-2
37	BWR	Stainless steel piping, piping components, and piping elements exposed to treated water >60°C (>140°F)	Cracking due to stress corrosion cracking, intergranular stress corrosion cracking	BWR Reactor Water Cleanup System	No	A-60	VII.E3-16

Table 3. Summary of Aging Management Programs for the Auxiliary Systems Evaluated in Chapter VII of the GALL Report

ID	Type	Component	Aging Effect/Mechanism	Aging Management Programs	Further Evaluation Recommended	Related Generic Item	Unique Item
38	BWR	Stainless steel piping, piping components, and piping elements exposed to treated water >60°C (>140°F)	Cracking due to stress corrosion cracking	BWR Stress Corrosion Cracking and Water Chemistry	No	A-61	VII.E4-15
39	BWR	Stainless steel BWR spent fuel storage racks exposed to treated water >60°C (>140°F)	Cracking due to stress corrosion cracking	Water Chemistry	No	A-96	VII.A2-6
40	BWR/PWR	Steel tanks in diesel fuel oil system exposed to air - outdoor (external)	Loss of material due to general, pitting, and crevice corrosion	Aboveground Steel Tanks	No	A-95	VII.H1-11
41	BWR/PWR	High-strength steel closure bolting exposed to air with steam or water leakage	Cracking due to cyclic loading, stress corrosion cracking	Bolting Integrity	No	A-04	VII.I-3
42	BWR/PWR	Steel closure bolting exposed to air with steam or water leakage	Loss of material due to general corrosion	Bolting Integrity	No	A-03	VII.I-6
43	BWR/PWR	Steel bolting and closure bolting exposed to air – indoor uncontrolled (external) or air – outdoor (External)	Loss of material due to general, pitting, and crevice corrosion	Bolting Integrity	No	AP-27 AP-28	VII.I-4 VII.I-1
44	BWR/PWR	Steel compressed air system closure bolting exposed to condensation	Loss of material due to general, pitting, and crevice corrosion	Bolting Integrity	No	A-103	VII.D-1
45	BWR/PWR	Steel closure bolting exposed to air – indoor uncontrolled (external)	Loss of preload due to thermal effects, gasket creep, and self-loosening	Bolting Integrity	No	AP-26	VII.I-5

Table 3. Summary of Aging Management Programs for the Auxiliary Systems Evaluated in Chapter VII of the GALL Report

ID	Type	Component	Aging Effect/Mechanism	Aging Management Programs	Further Evaluation Recommended	Related Generic Item	Unique Item
46	BWR/ PWR	Stainless steel and stainless clad steel piping, piping components, piping elements, and heat exchanger components exposed to closed cycle cooling water >60°C (>140°F)	Cracking due to stress corrosion cracking	Closed-Cycle Cooling Water System	No	A-68 AP-60	VII.E3-2 VII.C2-11 VII.E3-13 VII.E4-11
47	BWR/ PWR	Steel piping, piping components, piping elements, tanks, and heat exchanger components exposed to closed cycle cooling water	Loss of material due to general, pitting, and crevice corrosion	Closed-Cycle Cooling Water System	No	A-25	VII.C2-14 VII.F1-20 VII.F2-18 VII.F3-20 VII.F4-16 VII.H2-23
48	BWR/ PWR	Steel piping, piping components, piping elements, tanks, and heat exchanger components exposed to closed cycle cooling water	Loss of material due to general, pitting, crevice, and galvanic corrosion	Closed-Cycle Cooling Water System	No	A-63	VII.A3-3 VII.A4-3 VII.C2-1 VII.E1-6 VII.E3-4 VII.E4-2 VII.F1-11 VII.F2-9 VII.F3-11 VII.F4-8
49	BWR/ PWR	Stainless steel; steel with stainless steel cladding heat exchanger components exposed to closed cycle cooling water	Loss of material due to microbiologically influenced corrosion	Closed-Cycle Cooling Water System	No	A-67	VII.E3-1 VII.E4-1

Table 3. Summary of Aging Management Programs for the Auxiliary Systems Evaluated in Chapter VII of the GALL Report

ID	Type	Component	Aging Effect/Mechanism	Aging Management Programs	Further Evaluation Recommended	Related Generic Item	Unique Item
50	BWR/PWR	Stainless steel piping, piping components, and piping elements exposed to closed cycle cooling water	Loss of material due to pitting and crevice corrosion	Closed-Cycle Cooling Water System	No	A-52	VII.C2-10
51	BWR/PWR	Copper alloy piping, piping components, piping elements, and heat exchanger components exposed to closed cycle cooling water	Loss of material due to pitting, crevice, and galvanic corrosion	Closed-Cycle Cooling Water System	No	AP-12	VII.A3-5 VII.A4-6 VII.C2-4 VII.E1-11 VII.E3-8 VII.E4-5 VII.F1-15 VII.F2-13 VII.F3-15 VII.F4-11 VII.H1-2 VII.H2-8
						AP-34	VII.E1-2 VII.F1-8 VII.F3-8
52	BWR/PWR	Steel, stainless steel, and copper alloy heat exchanger tubes exposed to closed cycle cooling water	Reduction of heat transfer due to fouling	Closed-Cycle Cooling Water System	No	AP-63 AP-77	VII.C2-3 VII.E3-5 VII.E4-3 VII.F1-13 VII.F2-11 VII.F3-13 VII.F4-9
						AP-80	VII.C2-2 VII.F1-12 VII.F2-10 VII.F3-12

Table 3. Summary of Aging Management Programs for the Auxiliary Systems Evaluated in Chapter VII of the GALL Report

ID	Type	Component	Aging Effect/Mechanism	Aging Management Programs	Further Evaluation Recommended	Related Generic Item	Unique Item
53	BWR/ PWR	Steel compressed air system piping, piping components, and piping elements exposed to condensation (internal)	Loss of material due to general and pitting corrosion	Compressed Air Monitoring	No	A-26	VII.D-2
54	BWR/ PWR	Stainless steel compressed air system piping, piping components, and piping elements exposed to internal condensation	Loss of material due to pitting and crevice corrosion	Compressed Air Monitoring	No	AP-81	VII.D-4
55	BWR/ PWR	Steel ducting closure bolting exposed to air – indoor uncontrolled (external)	Loss of material due to general corrosion	External Surfaces Monitoring	No	A-105	VII.F1-4 VII.F2-4 VII.F3-4 VII.F4-3 VII.I-7
56	BWR/ PWR	Steel HVAC ducting and components external surfaces exposed to air – indoor uncontrolled (external)	Loss of material due to general corrosion	External Surfaces Monitoring	No	A-10	VII.F1-2 VII.F2-2 VII.F3-2 VII.F4-1
57	BWR/ PWR	Steel piping and components external surfaces exposed to air – indoor uncontrolled (External)	Loss of material due to general corrosion	External Surfaces Monitoring	No	A-80	VII.D-3
58	BWR/ PWR	Steel external surfaces exposed to air – indoor uncontrolled (external), air - outdoor (external), and condensation (external)	Loss of material due to general corrosion	External Surfaces Monitoring	No	A-77 A-78 A-81	VII.I-8 VII.I-9 VII.I-11

Table 3. Summary of Aging Management Programs for the Auxiliary Systems Evaluated in Chapter VII of the GALL Report

ID	Type	Component	Aging Effect/Mechanism	Aging Management Programs	Further Evaluation Recommended	Related Generic Item	Unique Item
59	BWR/PWR	Steel heat exchanger components exposed to air – indoor uncontrolled (external) or air -outdoor (external)	Loss of material due to general, pitting, and crevice corrosion	External Surfaces Monitoring	No	AP-40 AP-41	VII.G-6 VII.H2-4 VII.F1-10 VII.F2-8 VII.F3-10 VII.F4-7 VII.G-5 VII.H2-3
60	BWR/PWR	Steel piping, piping components, and piping elements exposed to air - outdoor (external)	Loss of material due to general, pitting, and crevice corrosion	External Surfaces Monitoring	No	A-24	VII.H1-8
61	BWR/PWR	Elastomer fire barrier penetration seals exposed to air – outdoor or air – indoor uncontrolled	Increased hardness, shrinkage and loss of strength due to weathering	Fire Protection	No	A-19 A-20	VII.G-1 VII.G-2
62	BWR/PWR	Aluminum piping, piping components, and piping elements exposed to raw water	Loss of material due to pitting and crevice corrosion	Fire Protection	No	AP-83	VII.G-8
63	BWR/PWR	Steel fire rated doors exposed to air – outdoor or air – indoor uncontrolled	Loss of material due to Wear	Fire Protection	No	A-21 A-22	VII.G-3 VII.G-4
64	BWR/PWR	Steel piping, piping components, and piping elements exposed to fuel oil	Loss of material due to general, pitting, and crevice corrosion	Fire Protection and Fuel Oil Chemistry	No	A-28	VII.G-21
65	BWR/PWR	Reinforced concrete structural fire barriers – walls, ceilings and floors exposed to air – indoor uncontrolled	Concrete cracking and spalling due to aggressive chemical attack, and reaction with aggregates	Fire Protection and Structures Monitoring Program	No	A-90	VII.G-28

Table 3. Summary of Aging Management Programs for the Auxiliary Systems Evaluated in Chapter VII of the GALL Report

ID	Type	Component	Aging Effect/Mechanism	Aging Management Programs	Further Evaluation Recommended	Related Generic Item	Unique Item
66	BWR/ PWR	Reinforced concrete structural fire barriers – walls, ceilings and floors exposed to air – outdoor	Concrete cracking and spalling due to freeze thaw, aggressive chemical attack, and reaction with aggregates	Fire Protection and Structures Monitoring Program	No	A-92	VII.G-30
67	BWR/ PWR	Reinforced concrete structural fire barriers – walls, ceilings and floors exposed to air – outdoor or air - indoor uncontrolled	Loss of material due to corrosion of embedded steel	Fire Protection and Structures Monitoring Program	No	A-91 A-93	VII.G-29 VII.G-31
68	BWR/ PWR	Steel piping, piping components, and piping elements exposed to raw water	Loss of material due to general, pitting, crevice, and microbiologically influenced corrosion, and fouling	Fire Water System	No	A-33	VII.G-24
69	BWR/ PWR	Stainless steel piping, piping components, and piping elements exposed to raw water	Loss of material due to pitting and crevice corrosion, and fouling	Fire Water System	No	A-55	VII.G-19
70	BWR/ PWR	Copper alloy piping, piping components, and piping elements exposed to raw water	Loss of material due to pitting, crevice, and microbiologically influenced corrosion, and fouling	Fire Water System	No	A-45	VII.G-12
71	BWR/ PWR	Steel piping, piping components, and piping elements exposed to moist air or condensation (Internal)	Loss of material due to general, pitting, and crevice corrosion	Inspection of Internal Surfaces in Miscellaneous Piping and Ducting Components	No	A-23	VII.G-23 VII.H2-21

Table 3. Summary of Aging Management Programs for the Auxiliary Systems Evaluated in Chapter VII of the GALL Report

ID	Type	Component	Aging Effect/Mechanism	Aging Management Programs	Further Evaluation Recommended	Related Generic Item	Unique Item
72	BWR/ PWR	Steel HVAC ducting and components internal surfaces exposed to condensation (Internal)	Loss of material due to general, pitting, crevice, and (for drip pans and drain lines) microbiologically influenced corrosion	Inspection of Internal Surfaces in Miscellaneous Piping and Ducting Components	No	A-08	VII.F1-3 VII.F2-3 VII.F3-3 VII.F4-2
73	BWR/ PWR	Steel crane structural girders in load handling system exposed to air – indoor uncontrolled (external)	Loss of material due to general corrosion	Inspection of Overhead Heavy Load and Light Load (Related to Refueling) Handling Systems	No	A-07	VII.B-3
74	BWR/ PWR	Steel cranes - rails exposed to air – indoor uncontrolled (external)	Loss of material due to Wear	Inspection of Overhead Heavy Load and Light Load (Related to Refueling) Handling Systems	No	A-05	VII.B-1
75	BWR/ PWR	Elastomer seals and components exposed to raw water	Hardening and loss of strength due to elastomer degradation; loss of material due to erosion	Open-Cycle Cooling Water System	No	AP-75 AP-76	VII.C1-1 VII.C1-2
76	BWR/ PWR	Steel piping, piping components, and piping elements (without lining/coating or with degraded lining/coating) exposed to raw water	Loss of material due to general, pitting, crevice, and microbiologically influenced corrosion, fouling, and lining/coating degradation	Open-Cycle Cooling Water System	No	A-38	VII.C1-19 VII.C3-10 VII.H2-22

Table 3. Summary of Aging Management Programs for the Auxiliary Systems Evaluated in Chapter VII of the GALL Report

ID	Type	Component	Aging Effect/Mechanism	Aging Management Programs	Further Evaluation Recommended	Related Generic Item	Unique Item
77	BWR/PWR	Steel heat exchanger components exposed to raw water	Loss of material due to general, pitting, crevice, galvanic, and microbiologically influenced corrosion, and fouling	Open-Cycle Cooling Water System	No	A-64	VII.C1-5
78	BWR/PWR	Stainless steel, nickel alloy, and copper alloy piping, piping components, and piping elements exposed to raw water	Loss of material due to pitting and crevice corrosion	Open-Cycle Cooling Water System	No	A-43 A-53 AP-53	VII.C3-2 VII.C3-7 VII.C1-13 VII.C3-6
79	BWR/PWR	Stainless steel piping, piping components, and piping elements exposed to raw water	Loss of material due to pitting and crevice corrosion, and fouling	Open-Cycle Cooling Water System	No	A-54	VII.C1-15
80	BWR/PWR	Stainless steel and copper alloy piping, piping components, and piping elements exposed to raw water	Loss of material due to pitting, crevice, and microbiologically influenced corrosion	Open-Cycle Cooling Water System	No	AP-45 AP-55	VII.H2-11 VII.H2-18
81	BWR/PWR	Copper alloy piping, piping components, and piping elements, exposed to raw water	Loss of material due to pitting, crevice, and microbiologically influenced corrosion, and fouling	Open-Cycle Cooling Water System	No	A-44	VII.C1-9
82	BWR/PWR	Copper alloy heat exchanger components exposed to raw water	Loss of material due to pitting, crevice, galvanic, and microbiologically influenced corrosion, and fouling	Open-Cycle Cooling Water System	No	A-65	VII.C1-3

Table 3. Summary of Aging Management Programs for the Auxiliary Systems Evaluated in Chapter VII of the GALL Report

ID	Type	Component	Aging Effect/Mechanism	Aging Management Programs	Further Evaluation Recommended	Related Generic Item	Unique Item
83	BWR/ PWR	Stainless steel and copper alloy heat exchanger tubes exposed to raw water	Reduction of heat transfer due to fouling	Open-Cycle Cooling Water System	No	A-72 AP-61	VII.C1-6 VII.C1-7 VII.C3-1 VII.G-7 VII.H2-6
84	BWR/ PWR	Copper alloy >15% Zn piping, piping components, piping elements, and heat exchanger components exposed to raw water, treated water, or closed cycle cooling water	Loss of material due to selective leaching	Selective Leaching of Materials	No	A-47 A-66 AP-32 AP-43 AP-65	VII.C1-10 VII.C3-3 VII.G-13 VII.H2-13 VII.C1-4 VII.A4-9 VII.C2-7 VII.E3-11 VII.E4-9 VII.A3-6 VII.A4-8 VII.C2-6 VII.E1-13 VII.E3-10 VII.E4-8 VII.F1-17 VII.F2-15 VII.F3-17 VII.F4-13 VII.H1-4 VII.H2-12 VII.E1-3 VII.F1-9 VII.F3-9

Table 3. Summary of Aging Management Programs for the Auxiliary Systems Evaluated in Chapter VII of the GALL Report

ID	Type	Component	Aging Effect/Mechanism	Aging Management Programs	Further Evaluation Recommended	Related Generic Item	Unique Item
85	BWR/PWR	Gray cast iron piping, piping components, and piping elements exposed to soil, raw water, treated water, or closed-cycle cooling water	Loss of material due to selective leaching	Selective Leaching of Materials	No	A-02 A-50 A-51 AP-31	VII.C1-12 VII.C3-5 VII.G-15 VII.H1-5 VII.H2-15 VII.C2-8 VII.F3-18 VII.C1-11 VII.C3-4 VII.G-14 VII.H2-14 VII.A3-7 VII.A4-10 VII.C2-9 VII.E1-14 VII.E3-12 VII.E4-10 VII.F1-18 VII.F2-16 VII.F4-14 VII.G-16
86	BWR/PWR	Structural steel (new fuel storage rack assembly) exposed to air – indoor uncontrolled (external)	Loss of material due to general, pitting, and crevice corrosion	Structures Monitoring Program	No	A-94	VII.A1-1
87	PWR	Boraflex spent fuel storage racks neutron-absorbing sheets exposed to treated borated water	Reduction of neutron-absorbing capacity due to boraflex degradation	Boraflex Monitoring	No	A-86	VII.A2-4

Table 3. Summary of Aging Management Programs for the Auxiliary Systems Evaluated in Chapter VII of the GALL Report

ID	Type	Component	Aging Effect/Mechanism	Aging Management Programs	Further Evaluation Recommended	Related Generic Item	Unique Item
88	PWR	Aluminum and copper alloy >15% Zn piping, piping components, and piping elements exposed to air with borated water leakage	Loss of material due to Boric acid corrosion	Boric Acid Corrosion	No	AP-1 AP-66	VII.A3-4 VII.E1-10 VII.I-12
89	PWR	Steel bolting and external surfaces exposed to air with borated water leakage	Loss of material due to Boric acid corrosion	Boric Acid Corrosion	No	A-79 A-102	VII.A3-2 VII.E1-1 VII.I-10 VII.I-2
90	PWR	Stainless steel and steel with stainless steel cladding piping, piping components, piping elements, tanks, and fuel storage racks exposed to treated borated water >60°C (>140°F)	Cracking due to stress corrosion cracking	Water Chemistry	No	A-56 A-97 AP-82	VII.A3-10 VII.A2-7 VII.E1-20
91	PWR	Stainless steel and steel with stainless steel cladding piping, piping components, and piping elements exposed to treated borated water	Loss of material due to pitting and crevice corrosion	Water Chemistry	No	AP-79	VII.A2-1 VII.A3-8 VII.E1-17
92	BWR/ PWR	Galvanized steel piping, piping components, and piping elements exposed to air – indoor uncontrolled	None	None	NA - No AEM or AMP	AP-13	VII.J-6
93	BWR/ PWR	Glass piping elements exposed to air, air – indoor uncontrolled (external), fuel oil, lubricating oil, raw water, treated water, and treated borated water	None	None	NA - No AEM or AMP	AP-14 AP-15 AP-48 AP-49 AP-50 AP-51 AP-52	VII.J-8 VII.J-10 VII.J-7 VII.J-9 VII.J-11 VII.J-13 VII.J-12

Table 3. Summary of Aging Management Programs for the Auxiliary Systems Evaluated in Chapter VII of the GALL Report

ID	Type	Component	Aging Effect/Mechanism	Aging Management Programs	Further Evaluation Recommended	Related Generic Item	Unique Item
94	BWR/ PWR	Stainless steel and nickel alloy piping, piping components, and piping elements exposed to air – indoor uncontrolled (external)	None	None	NA - No AEM or AMP	AP-16 AP-17	VII.J-14 VII.J-15
95	BWR/ PWR	Steel and aluminum piping, piping components, and piping elements exposed to air – indoor controlled (external)	None	None	NA - No AEM or AMP	AP-2 AP-36	VII.J-20 VII.J-1
96	BWR/ PWR	Steel and stainless steel piping, piping components, and piping elements in concrete	None	None	NA - No AEM or AMP	AP-3 AP-19	VII.J-21 VII.J-17
97	BWR/ PWR	Steel, stainless steel, aluminum, and copper alloy piping, piping components, and piping elements exposed to gas	None	None	NA - No AEM or AMP	AP-6 AP-9 AP-22 AP-37	VII.J-23 VII.J-4 VII.J-19 VII.J-2
98	BWR/ PWR	Steel, stainless steel, and copper alloy piping, piping components, and piping elements exposed to dried air	None	None	NA - No AEM or AMP	AP-4 AP-8 AP-20	VII.J-22 VII.J-3 VII.J-18
99	PWR	Stainless steel and copper alloy <15% Zn piping, piping components, and piping elements exposed to air with borated water leakage	None	None	NA - No AEM or AMP	AP-11 AP-18	VII.J-5 VII.J-16

Table 4. Summary of Aging Management Programs for the Steam and Power Conversion System Evaluated in Chapter VIII of the GALL Report

ID	Type	Component	Aging Effect/Mechanism	Aging Management Programs	Further Evaluation Recommended	Related Generic Item	Unique Item
1	BWR/PWR	Steel piping, piping components, and piping elements exposed to steam or treated water	Cumulative fatigue damage	TLAA, evaluated in accordance with 10 CFR 54.21(c)	Yes, TLAA	S-08 S-11	VIII.B1-10 VIII.B2-5 VIII.D1-7 VIII.D2-6 VIII.G-37
2	BWR/PWR	Steel piping, piping components, and piping elements exposed to steam	Loss of material due to general, pitting and crevice corrosion	Water Chemistry and One-Time Inspection	Yes, detection of aging effects is to be evaluated	S-04 S-06	VIII.A-15 VIII.C-3 VIII.A-16 VIII.C-4
3	PWR	Steel heat exchanger components exposed to treated water	Loss of material due to general, pitting and crevice corrosion	Water Chemistry and One-Time Inspection	Yes, detection of aging effects is to be evaluated	S-19	VIII.E-37 VIII.F-28
4	BWR/PWR	Steel piping, piping components, and piping elements exposed to treated water	Loss of material due to general, pitting and crevice corrosion	Water Chemistry and One-Time Inspection	Yes, detection of aging effects is to be evaluated	S-09 S-10	VIII.B2-6 VIII.C-6 VIII.D2-7 VIII.E-33 VIII.B1-11 VIII.C-7 VIII.D1-8 VIII.E-34 VIII.F-25 VIII.G-38
5	BWR	Steel heat exchanger components exposed to treated water	Loss of material due to general, pitting, crevice, and galvanic corrosion	Water Chemistry and One-Time Inspection	Yes, detection of aging effects is to be evaluated	S-18	VIII.E-7

Table 4. Summary of Aging Management Programs for the Steam and Power Conversion System Evaluated in Chapter VIII of the GALL Report

ID	Type	Component	Aging Effect/Mechanism	Aging Management Programs	Further Evaluation Recommended	Related Generic Item	Unique Item
6	BWR/ PWR	Steel and stainless steel tanks exposed to treated water	Loss of material due to general (steel only) pitting and crevice corrosion	Water Chemistry and One-Time Inspection	Yes, detection of aging effects is to be evaluated	S-13	VIII.E-40 VIII.G-41
7	BWR/ PWR	Steel piping, piping components, and piping elements exposed to lubricating oil	Loss of material due to general, pitting and crevice corrosion	Lubricating Oil Analysis and One-Time Inspection	Yes, detection of aging effects is to be evaluated	SP-25	VIII.A-14 VIII.D1-6 VIII.D2-5 VIII.E-32 VIII.G-35
8	BWR/ PWR	Steel piping, piping components, and piping elements exposed to raw water	Loss of material due to general, pitting, crevice, and microbiologically-influenced corrosion, and fouling	Plant specific	Yes, plant specific	S-12	VIII.G-36
9	BWR/ PWR	Stainless steel and copper alloy heat exchanger tubes exposed to treated water	Reduction of heat transfer due to fouling	Water Chemistry and One-Time Inspection	Yes, detection of aging effects is to be evaluated	SP-40 SP-58	VIII.E-13 VIII.F-10 VIII.E-10 VIII.F-7 VIII.G-10
10	BWR/ PWR	Steel, stainless steel, and copper alloy heat exchanger tubes exposed to lubricating oil	Reduction of heat transfer due to fouling	Lubricating Oil Analysis and One-Time Inspection	Yes, detection of aging effects is to be evaluated	SP-53 SP-62 SP-63	VIII.G-8 VIII.G-12 VIII.G-15

Table 4. Summary of Aging Management Programs for the Steam and Power Conversion System Evaluated in Chapter VIII of the GALL Report

ID	Type	Component	Aging Effect/Mechanism	Aging Management Programs	Further Evaluation Recommended	Related Generic Item	Unique Item
11	BWR/ PWR	Buried steel piping, piping components, piping elements, and tanks (with or without coating or wrapping) exposed to soil	Loss of material due to general, pitting, crevice, and microbiologically-influenced corrosion	Buried Piping and Tanks Surveillance or Buried Piping and Tanks Inspection	No Yes, detection of aging effects and operating experience are to be further evaluated	S-01	VIII.E-1 VIII.G-1
12	BWR/ PWR	Steel heat exchanger components exposed to lubricating oil	Loss of material due to general, pitting, crevice, and microbiologically-influenced corrosion	Lubricating Oil Analysis and One-Time Inspection	Yes, detection of aging effects is to be evaluated	S-17	VIII.G-6
13	BWR	Stainless steel piping, piping components, piping elements exposed to steam	Cracking due to stress corrosion cracking	Water Chemistry and One-Time Inspection	Yes, detection of aging effects is to be evaluated	SP-45	VIII.A-11 VIII.B2-1
14	BWR/ PWR	Stainless steel piping, piping components, piping elements, tanks, and heat exchanger components exposed to treated water >60°C (>140°F)	Cracking due to stress corrosion cracking	Water Chemistry and One-Time Inspection	Yes, detection of aging effects is to be evaluated	S-39 SP-17 SP-19 SP-42	VIII.F-3 VIII.B1-5 VIII.C-2 VIII.D1-5 VIII.E-30 VIII.F-24 VIII.G-33 VIII.E-31 VIII.E-38

Table 4. Summary of Aging Management Programs for the Steam and Power Conversion System Evaluated in Chapter VIII of the GALL Report

ID	Type	Component	Aging Effect/Mechanism	Aging Management Programs	Further Evaluation Recommended	Related Generic Item	Unique Item
15	BWR/PWR	Aluminum and copper alloy piping, piping components, and piping elements exposed to treated water	Loss of material due to pitting and crevice corrosion	Water Chemistry and One-Time Inspection	Yes, detection of aging effects is to be evaluated	SP-24 / SP-61	VIII.D1-1, VIII.D2-1, VIII.E-15, VIII.F-12, VIII.G-17, VIII.A-5, VIII.F-15
16	BWR/PWR	Stainless steel piping, piping components, and piping elements; tanks, and heat exchanger components exposed to treated water	Loss of material due to pitting and crevice corrosion	Water Chemistry and One-Time Inspection	Yes, detection of aging effects is to be evaluated	S-21, S-22, SP-16	VIII.E-4, VIII.E-36, VIII.F-27, VIII.B1-4, VIII.C-1, VIII.D1-4, VIII.D2-4, VIII.E-29, VIII.F-23, VIII.G-32
17	BWR/PWR	Stainless steel piping, piping components, and piping elements exposed to soil	Loss of material due to pitting and crevice corrosion	Plant specific	Yes, plant specific	SP-37	VIII.E-28, VIII.G-31
18	BWR/PWR	Copper alloy piping, piping components, and piping elements exposed to lubricating oil	Loss of material due to pitting and crevice corrosion	Lubricating Oil Analysis and One-Time Inspection	Yes, detection of aging effects is to be evaluated	SP-32	VIII.A-3, VIII.D1-2, VIII.D2-2, VIII.E-17, VIII.G-19

Table 4. Summary of Aging Management Programs for the Steam and Power Conversion System Evaluated in Chapter VIII of the GALL Report

ID	Type	Component	Aging Effect/Mechanism	Aging Management Programs	Further Evaluation Recommended	Related Generic Item	Unique Item
19	BWR/ PWR	Stainless steel piping, piping components, piping elements, and heat exchanger components exposed to lubricating oil	Loss of material due to pitting, crevice, and microbiologically-influenced corrosion	Lubricating Oil Analysis and One-Time Inspection	Yes, detection of aging effects is to be evaluated	S-20 SP-38	VIII.G-3 VIII.A-9 VIII.D1-3 VIII.D2-3 VIII.E-26 VIII.G-29
20	BWR/ PWR	Steel tanks exposed to air – outdoor (external)	Loss of material/ general, pitting, and crevice corrosion	Aboveground Steel Tanks	No	S-31	VIII.E-39 VIII.G-40
21	BWR/ PWR	High-strength steel closure bolting exposed to air with steam or water leakage	Cracking due to cyclic loading, stress corrosion cracking	Bolting Integrity	No	S-03	VIII.H-3
22	BWR/ PWR	Steel bolting and closure bolting exposed to air with steam or water leakage, air – outdoor (external), or air – indoor uncontrolled (external);	Loss of material due to general, pitting and crevice corrosion; loss of preload due to thermal effects, gasket creep, and self-loosening	Bolting Integrity	No	S-02 S-32 S-33 S-34	VIII.H-6 VIII.H-1 VIII.H-5 VIII.H-4
23	BWR/ PWR	Stainless steel piping, piping components, and piping elements exposed to closed-cycle cooling water >60°C (>140°F)	Cracking due to stress corrosion cracking	Closed-Cycle Cooling Water System	No	SP-54	VIII.E-25 VIII.F-21 VIII.G-28
24	BWR/ PWR	Steel heat exchanger components exposed to closed cycle cooling water	Loss of material due to general, pitting, crevice, and galvanic corrosion	Closed-Cycle Cooling Water System	No	S-23	VIII.A-1 VIII.E-5 VIII.F-4 VIII.G-5

Table 4. Summary of Aging Management Programs for the Steam and Power Conversion System Evaluated in Chapter VIII of the GALL Report

ID	Type	Component	Aging Effect/Mechanism	Aging Management Programs	Further Evaluation Recommended	Related Generic Item	Unique Item
25	BWR/ PWR	Stainless steel piping, piping components, piping elements, and heat exchanger components exposed to closed cycle cooling water	Loss of material due to pitting and crevice corrosion	Closed-Cycle Cooling Water System	No	S-25 SP-39	VIII.E-2 VIII.F-1 VIII.G-2 VIII.E-24 VIII.F-20 VIII.G-27
26	BWR/ PWR	Copper alloy piping, piping components, and piping elements exposed to closed cycle cooling water	Loss of material due to pitting, crevice, and galvanic corrosion	Closed-Cycle Cooling Water System	No	SP-8	VIII.E-16 VIII.F-13 VIII.G-18
27	BWR/ PWR	Steel, stainless steel, and copper alloy heat exchanger tubes exposed to closed cycle cooling water	Reduction of heat transfer due to fouling	Closed-Cycle Cooling Water System	No	SP-41 SP-57 SP-64	VIII.E-11 VIII.F-8 VIII.G-11 VIII.E-8 VIII.A-2 VIII.E-14 VIII.F-11 VIII.G-14
28	BWR/ PWR	Steel external surfaces exposed to air – indoor uncontrolled (external), condensation (external), or air outdoor (external)	Loss of material due to general corrosion	External Surfaces Monitoring	No	S-29 S-41 S-42	VIII.H-7 VIII.H-8 VIII.H-10

Table 4. Summary of Aging Management Programs for the Steam and Power Conversion System Evaluated in Chapter VIII of the GALL Report

ID	Type	Component	Aging Effect/Mechanism	Aging Management Programs	Further Evaluation Recommended	Related Generic Item	Unique Item
29	BWR/ PWR	Steel piping, piping components, and piping elements exposed to steam or treated water	Wall thinning due to flow-accelerated corrosion	Flow-Accelerated Corrosion	No	S-15 S-16	VIII.A-17 VIII.B1-9 VIII.B2-4 VIII.C-5 VIII.D1-9 VIII.D2-8 VIII.E-35 VIII.F-26 VIII.G-39
30	BWR/ PWR	Steel piping, piping components, and piping elements exposed to air outdoor (internal) or condensation (internal)	Loss of material due to general, pitting, and crevice corrosion	Inspection of Internal Surfaces in Miscellaneous Piping and Ducting Components	No	SP-59 SP-60	VIII.B1-6 VIII.B1-7 VIII.G-34
31	BWR/ PWR	Steel heat exchanger components exposed to raw water	Loss of material due to general, pitting, crevice, galvanic, and microbiologically-influenced corrosion, and fouling	Open-Cycle Cooling Water System	No	S-24	VIII.E-6 VIII.F-5 VIII.G-7
32	BWR/ PWR	Stainless steel and copper alloy piping, piping components, and piping elements exposed to raw water	Loss of material due to pitting, crevice, and microbiologically-influenced corrosion	Open-Cycle Cooling Water System	No	SP-31 SP-36	VIII.A-4 VIII.E-18 VIII.F-14 VIII.G-20 VIII.E-27 VIII.F-22 VIII.G-30

Table 4. Summary of Aging Management Programs for the Steam and Power Conversion System Evaluated in Chapter VIII of the GALL Report

ID	Type	Component	Aging Effect/Mechanism	Aging Management Programs	Further Evaluation Recommended	Related Generic Item	Unique Item
33	BWR/PWR	Stainless steel heat exchanger components exposed to raw water	Loss of material due to pitting, crevice, and microbiologically-influenced corrosion, and fouling	Open-Cycle Cooling Water System	No	S-26	VIII.E-3 VIII.F-2 VIII.G-4
34	BWR/PWR	Steel, stainless steel, and copper alloy heat exchanger tubes exposed to raw water	Reduction of heat transfer due to fouling	Open-Cycle Cooling Water System	No	S-27 S-28 SP-56	VIII.G-16 VIII.E-12 VIII.F-9 VIII.G-13 VIII.E-9 VIII.F-6 VIII.G-9
35	BWR/PWR	Copper alloy >15% Zn piping, piping components, and piping elements exposed to closed cycle cooling water, raw water, or treated water	Loss of material due to selective leaching	Selective Leaching of Materials	No	SP-29 SP-30 SP-55	VIII.E-19 VIII.F-16 VIII.G-21 VIII.A-6 VIII.E-20 VIII.F-17 VIII.G-22 VIII.E-21 VIII.F-18 VIII.G-23

Table 4. Summary of Aging Management Programs for the Steam and Power Conversion System Evaluated in Chapter VIII of the GALL Report

ID	Type	Component	Aging Effect/Mechanism	Aging Management Programs	Further Evaluation Recommended	Related Generic Item	Unique Item
36	BWR/ PWR	Gray cast iron piping, piping components, and piping elements exposed to soil, treated water, or raw water	Loss of material due to selective leaching	Selective Leaching of Materials	No	SP-26 SP-27 SP-28	VIII.E-22 VIII.G-25 VIII.A-8 VIII.E-23 VIII.F-19 VIII.G-26 VIII.A-7 VIII.G-24
37	BWR/ PWR	Steel, stainless steel, and nickel-based alloy piping, piping components, and piping elements exposed to steam	Loss of material due to pitting and crevice corrosion	Water Chemistry	No	S-05 S-07 SP-18 SP-43 SP-46	VIII.B2-3 VIII.B1-8 VIII.B1-1 VIII.A-12 VIII.B1-3 VIII.A-13 VIII.B2-2
38	PWR	Steel bolting and external surfaces exposed to air with borated water leakage	Loss of material due to boric acid corrosion	Boric Acid Corrosion	No	S-30 S-40	VIII.H-9 VIII.H-2
39	PWR	Stainless steel piping, piping components, and piping elements exposed to steam	Cracking due to stress corrosion cracking	Water Chemistry	No	SP-44	VIII.A-10 VIII.B1-2
40	BWR/ PWR	Glass piping elements exposed to air, lubricating oil, raw water, and treated water	None	None	NA - No AEM or AMP	SP-9 SP-10 SP-33 SP-34 SP-35	VIII.I-5 VIII.I-6 VIII.I-4 VIII.I-7 VIII.I-8

Table 4. Summary of Aging Management Programs for the Steam and Power Conversion System Evaluated in Chapter VIII of the GALL Report

ID	Type	Component	Aging Effect/Mechanism	Aging Management Programs	Further Evaluation Recommended	Related Generic Item	Unique Item
41	BWR/ PWR	Stainless steel, copper alloy, and nickel alloy piping, piping components, and piping elements exposed to air – indoor uncontrolled (external)	None	None	NA - No AEM or AMP	SP-6 SP-11 SP-12	VIII.I-2 VIII.I-9 VIII.I-10
42	BWR/ PWR	Steel piping, piping components, and piping elements exposed to air – indoor controlled (external)	None	None	NA - No AEM or AMP	SP-1	VIII.I-13
43	BWR/ PWR	Steel and stainless steel piping, piping components, and piping elements in concrete	None	None	NA - No AEM or AMP	SP-2 SP-13	VIII.I-14 VIII.I-11
44	BWR/ PWR	Steel, stainless steel, aluminum, and copper alloy piping, piping components, and piping elements exposed to gas	None	None	NA - No AEM or AMP	SP-4 SP-5 SP-15 SP-23	VIII.I-15 VIII.I-3 VIII.I-12 VIII.I-1

Table 5. Summary of Aging Management Programs for Structures and Component Supports Evaluated in Chapters II and III of the GALL Report

PWR Concrete (Reinforced and Prestressed) and Steel Containment

BWR Concrete (Mark II and III) and Steel (Mark I, II, and III) Containment

ID	Type	Component	Aging Effect/Mechanism	Aging Management Programs	Further Evaluation Recommended	Related Generic Item	Unique Item
1	BWR/ PWR	Concrete elements: walls, dome, basemat, ring girder, buttresses, containment (as applicable).	Aging of accessible and inaccessible concrete areas due to aggressive chemical attack, and corrosion of embedded steel	ISI (IWL) and for inaccessible concrete, an examination of representative samples of below-grade concrete and periodic monitoring of groundwater if environment is non-aggressive. A plant specific program is to be evaluated if environment is aggressive.	Yes, plant-specific, if the environment is aggressive	C-03 C-05 C-25 C-26 C-27 C-41 C-42 C-43	II.A1-4 II.A1-7 II.A2-4 II.B3.1-1 II.B1.2-5 II.B2.2-5 II.B3.2-5 II.B1.2-2 II.B2.2-2 II.B3.2-7 II.A2-7 II.B3.1-6
2	BWR/ PWR	Concrete elements; All	Cracks and distortion due to increased stress levels from settlement	Structures Monitoring Program. If a de-watering system is relied upon for control of settlement, then the licensee is to ensure proper functioning of the de-watering system through the period of extended operation.	Yes, if not within the scope of the applicant's structures monitoring program or a de-watering system is relied upon	C-06 C-36 C-37	II.B1.2-1 II.B2.2-1 II.B3.2-1 II.A2-5 II.B3.1-2 II.A1-5
3	BWR/ PWR	Concrete elements: foundation, sub-foundation	Reduction in foundation strength, cracking, differential settlement due to erosion of porous concrete subfoundation	Structures Monitoring Program. If a de-watering system is relied upon to control erosion of cement from porous concrete subfoundations, then the licensee is to ensure proper functioning of the de-watering system through the period of extended operation.	Yes, if not within the scope of the applicant's structures monitoring program or a de-watering system is relied upon	C-07	II.A1-8 II.A2-8 II.B1.2-7 II.B2.2-7 II.B3.1-7 II.B3.2-8

Table 5. Summary of Aging Management Programs for Structures and Component Supports Evaluated in Chapters II and III of the GALL Report

ID	Type	Component	Aging Effect/Mechanism	Aging Management Programs	Further Evaluation Recommended	Related Generic Item	Unique Item
4	BWR/ PWR	Concrete elements: dome, wall, basemat, ring girder, buttresses, containment, concrete fill-in annulus (as applicable)	Reduction of strength and modulus due to elevated temperature	Plant-specific	Yes, plant-specific if temperature limits are exceeded	C-08 C-33 C-34 C-35 C-50	II.A1-1 II.B3.2-2 II.A2-1 II.B1.2-3 II.B2.2-3 II.B3.1-4
5	BWR	Steel elements: Drywell; torus; drywell head; embedded shell and sand pocket regions; drywell support skirt; torus ring girder; downcomers; liner plate, ECCS suction header, support skirt, region shielded by diaphragm floor, suppression chamber (as applicable)	Loss of material due to general, pitting and crevice corrosion	ISI (IWE) and 10 CFR Part 50, Appendix J	Yes, if corrosion is significant for inaccessible areas	C-19 C-46	II.B1.1-2 II.B3.1-8 II.B1.2-8 II.B2.1-1 II.B2.2-10
6	BWR/ PWR	Steel elements: steel liner, liner anchors, integral attachments	Loss of material due to general, pitting and crevice corrosion	ISI (IWE) and 10 CFR Part 50, Appendix J	Yes, if corrosion is significant for inaccessible areas	C-09	II.A1-11 II.A2-9 II.B3.2-9
7	BWR/ PWR	Prestressed containment tendons	Loss of prestress due to relaxation, shrinkage, creep, and elevated temperature	TLAA evaluated in accordance with 10 CFR 54.21(c)	Yes, TLAA	C-11	II.A1-9 II.B2.2-8

Table 5. Summary of Aging Management Programs for Structures and Component Supports Evaluated in Chapters II and III of the GALL Report

ID	Type	Component	Aging Effect/Mechanism	Aging Management Programs	Further Evaluation Recommended	Related Generic Item	Unique Item
8	BWR	Steel and stainless steel elements: vent line, vent header, vent line bellows; downcomers;	Cumulative fatigue damage (CLB fatigue analysis exists)	TLAA evaluated in accordance with 10 CFR 54.21(c)	Yes, TLAA	C-21 C-48	II.B1.1-4 II.B2.2-14
9	BWR/ PWR	Steel, stainless steel elements, dissimilar metal welds: penetration sleeves, penetration bellows; suppression pool shell, unbraced downcomers	Cumulative fatigue damage (CLB fatigue analysis exists)	TLAA evaluated in accordance with 10 CFR 54.21(c)	Yes, TLAA	C-13 C-45	II.A3-4 II.B4-4 II.B2.1-4
10	BWR/ PWR	Stainless steel penetration sleeves, penetration bellows, dissimilar metal welds	Cracking due to stress corrosion cracking	ISI (IWE) and 10 CFR Part 50, Appendix J and additional appropriate examinations/evaluations for bellows assemblies and dissimilar metal welds	Yes, detection of aging effects is to be evaluated	C-15	II.A3-2 II.B4-2
11	BWR	Stainless steel vent line bellows,	Cracking due to stress corrosion cracking	ISI (IWE) and 10 CFR Part 50, Appendix J, and additional appropriate examination/evaluation for bellows assemblies and dissimilar metal welds	Yes, detection of aging effects is to be evaluated	C-22	II.B1.1-5

Table 5. Summary of Aging Management Programs for Structures and Component Supports Evaluated in Chapters II and III of the GALL Report

ID	Type	Component	Aging Effect/Mechanism	Aging Management Programs	Further Evaluation Recommended	Related Generic Item	Unique Item
12	BWR/PWR	Steel, stainless steel elements, dissimilar metal welds: penetration sleeves, penetration bellows; suppression pool shell, unbraced downcomers	Cracking due to cyclic loading	ISI (IWE) and 10 CFR Part 50, Appendix J supplemented to detect fine cracks	Yes, detection of aging effects is to be evaluated	C-14 C-44	II.A3-3 II.B4-3 II.B2.1-3
13	BWR	Steel, stainless steel elements, dissimilar metal welds: torus; vent line; vent header; vent line bellows; downcomers	Cracking due to cyclic loading	ISI (IWE) and 10 CFR Part 50, Appendix J supplemented to detect fine cracks	Yes, detection of aging effects is to be evaluated	C-20 C-47	II.B1.1-3 II.B2.2-13
14	BWR/PWR	Concrete elements: dome, wall, basemat ring girder, buttresses, containment (as applicable)	Loss of material (Scaling, cracking, and spalling) due to freeze-thaw	ISI (IWL) Evaluation is needed for plants that are located in moderate to severe weathering conditions (weathering index >100 day-inch/yr) (NUREG-1557).	Yes, for inaccessible areas of plants located in moderate to severe weathering conditions	C-01 C-28 C-29	II.A1-2 II.A2-2 II.B3.2-3
15	BWR/PWR	Concrete elements: walls, dome, basemat, ring girder, buttresses, containment, concrete fill-in annulus (as applicable).	Increase in porosity, permeability due to leaching of calcium hydroxide; cracking due to expansion and reaction with aggregate	ISI (IWL) for accessible areas. None for inaccessible areas if concrete was constructed in accordance with the recommendations in ACI 201.2R.	Yes, if concrete was not constructed as stated for inaccessible areas	C-02 C-04 C-30 C-31 C-32 C-38 C-39 C-40 C-51	II.A1-6 II.A1-3 II.A2-6 II.B3.1-3 II.B1.2-6 II.B2.2-6 II.B3.2-6 II.A2-3 II.B1.2-4 II.B2.2-4 II.B3.2-4 II.B3.1-5

ID	Type	Component	Aging Effect/Mechanism	Aging Management Programs	Further Evaluation Recommended	Related Generic Item	Unique Item
16	BWR/PWR	Seals, gaskets, and moisture barriers	Loss of sealing and leakage through containment due to deterioration of joint seals, gaskets, and moisture barriers (caulking, flashing, and other sealants)	ISI (IWE) and 10 CFR Part 50, Appendix J	No	C-18	II.A3-7 II.B4-7
17	BWR/PWR	Personnel airlock, equipment hatch and CRD hatch locks, hinges, and closure mechanisms	Loss of leak tightness in closed position due to mechanical wear of locks, hinges and closure mechanisms	10 CFR Part 50, Appendix J and Plant Technical Specifications	No	C-17	II.A3-5 II.B4-5
18	BWR/PWR	Steel penetration sleeves and dissimilar metal welds; personnel airlock, equipment hatch and CRD hatch	Loss of material due to general, pitting, and crevice corrosion	ISI (IWE) and 10 CFR Part 50, Appendix J	No	C-12 C-16	II.A3-1 II.B4-1 II.A3-6 II.B4-6
19	BWR	Steel elements: stainless steel suppression chamber shell (inner surface)	Cracking due to stress corrosion cracking	ISI (IWE) and 10 CFR Part 50, Appendix J	No	C-24	II.B3.1-9 II.B3.2-10
20	BWR	Steel elements: suppression chamber liner (interior surface)	Loss of material due to general, pitting, and crevice corrosion	ISI (IWE) and 10 CFR Part 50, Appendix J	No	C-49	II.B1.2-10 II.B2.2-12

Table 5. Summary of Aging Management Programs for Structures and Component Supports Evaluated in Chapters II and III of the GALL Report

ID	Type	Component	Aging Effect/Mechanism	Aging Management Programs	Further Evaluation Recommended	Related Generic Item	Unique Item
21	BWR	Steel elements: drywell head and downcomer pipes	Fretting or lock up due to mechanical wear	ISI (IWE)	No	C-23	II.B1.1-1 II.B1.2-9 II.B2.1-2 II.B2.2-11
22	BWR/ PWR	Prestressed containment: tendons and anchorage components	Loss of material due to corrosion	ISI (IWL)	No	C-10	II.A1-10 II.B2.2-9

Safety-Related and Other Structures; and Component Supports

ID	Type	Component	Aging Effect/Mechanism	Aging Management Programs	Further Evaluation Recommended	Related Generic Item	Unique Item
23	BWR/ PWR	All Groups except Group 6: interior and above grade exterior concrete	Cracking, loss of bond, and loss of material (spalling, scaling) due to corrosion of embedded steel	Structures Monitoring Program	Yes, if not within the scope of the applicant's structures monitoring program	T-01	III.A1-6 III.A2-6 III.A3-6 III.A5-6 III.A7-5 III.A8-5 III.A9-5
24	BWR/ PWR	All Groups except Group 6: interior and above grade exterior concrete	Increase in porosity and permeability, cracking, loss of material (spalling, scaling) due to aggressive chemical attack	Structures Monitoring Program	Yes, if not within the scope of the applicant's structures monitoring program	T-06	III.A1-10 III.A2-10 III.A3-10 III.A4-4 III.A5-10 III.A7-9 III.A9-9
25	BWR/ PWR	All Groups except Group 6: steel components: all structural steel	Loss of material due to corrosion	Structures Monitoring Program. If protective coatings are relied upon to manage the effects of aging, the structures monitoring program is to include provisions to address protective coating monitoring and maintenance.	Yes, if not within the scope of the applicant's structures monitoring program	T-11	III.A1-12 III.A2-12 III.A3-12 III.A4-5 III.A5-12 III.A7-10 III.A8-8

Table 5. Summary of Aging Management Programs for Structures and Component Supports Evaluated in Chapters II and III of the GALL Report

ID	Type	Component	Aging Effect/Mechanism	Aging Management Programs	Further Evaluation Recommended	Related Generic Item	Unique Item
26	BWR/ PWR	All Groups except Group 6: accessible and inaccessible concrete: foundation	Loss of material (spalling, scaling) and cracking due to freeze-thaw	Structures Monitoring Program. Evaluation is needed for plants that are located in moderate to severe weathering conditions (weathering index >100 day-inch/yr) (NUREG-1557).	Yes, if not within the scope of the applicant's structures monitoring program or for inaccessible areas of plants located in moderate to severe weathering conditions	T-01	III.A1-6 III.A2-6 III.A3-6 III.A5-6 III.A7-5 III.A8-5 III.A9-5
27	BWR/ PWR	All Groups except Group 6: accessible and inaccessible interior/exterior concrete	Cracking due to expansion due to reaction with aggregates	Structures Monitoring Program None for inaccessible areas if concrete was constructed in accordance with the recommendations in ACI 201.2R-77.	Yes, if not within the scope of the applicant's structures monitoring program or concrete was not constructed as stated for inaccessible areas	T-03	III.A1-2 III.A2-2 III.A3-2 III.A4-2 III.A5-2 III.A7-1 III.A8-1 III.A9-1
28	BWR/ PWR	Groups 1-3, 5-9: All	Cracks and distortion due to increased stress levels from settlement	Structures Monitoring Program. If a de-watering system is relied upon for control of settlement, then the licensee is to ensure proper functioning of the de-watering system through the period of extended operation.	Yes, if not within the scope of the applicant's structures monitoring program or a de-watering system is relied upon	T-08	III.A1-3 III.A2-3 III.A3-3 III.A5-3 III.A6-4 III.A7-2 III.A8-2 III.A9-2
29	BWR/ PWR	Groups 1-3, 5-9: foundation	Reduction in foundation strength, cracking, differential settlement due to erosion of porous concrete subfoundation	Structures Monitoring Program. If a de-watering system is relied upon for control of settlement, then the licensee is to ensure proper functioning of the de-watering system through the period of extended operation.	Yes, if not within the scope of the applicant's structures monitoring program or a de-watering system is relied upon	T-09	III.A1-8 III.A2-8 III.A3-8 III.A5-8 III.A6-8 III.A7-7 III.A8-7 III.A9-7

Table 5. Summary of Aging Management Programs for Structures and Component Supports Evaluated in Chapters II and III of the GALL Report

ID	Type	Component	Aging Effect/Mechanism	Aging Management Programs	Further Evaluation Recommended	Related Generic Item	Unique Item
30	BWR/ PWR	Group 4: Radial beam seats in BWR drywell; RPV support shoes for PWR with nozzle supports; Steam generator supports	Lock-up due to wear	ISI (IWF) or Structures Monitoring Program	Yes, if not within the scope of ISI or structures monitoring program	T-13	III.A4-6
31	BWR/ PWR	Groups 1-3, 5, 7-9: below-grade concrete components, such as exterior walls below grade and foundation	Increase in porosity and permeability, cracking, loss of material (spalling, scaling)/ aggressive chemical attack; Cracking, loss of bond, and loss of material (spalling, scaling)/ corrosion of embedded steel	Structures monitoring Program; Examination of representative samples of below-grade concrete, and periodic monitoring of groundwater, if the environment is non-aggressive. A plant specific program is to be evaluated if environment is aggressive.	Yes, plant-specific, if environment is aggressive	T-05 / T-07	III.A1-4 III.A2-4 III.A3-4 III.A5-4 III.A7-3 III.A8-3 III.A9-3 III.A1-5 III.A2-5 III.A3-5 III.A5-5 III.A7-4 III.A8-4 III.A9-4
32	BWR/ PWR	Groups 1-3, 5, 7-9: exterior above and below grade reinforced concrete foundations	Increase in porosity and permeability, loss of strength due to leaching of calcium hydroxide.	Structures Monitoring Program for accessible areas. None for inaccessible areas if concrete was constructed in accordance with the recommendations in ACI 201.2R-77.	Yes, if concrete was not constructed as stated for inaccessible areas	T-02	III.A1-7 III.A2-7 III.A3-7 III.A5-7 III.A7-6 III.A8-6 III.A9-6

Table 5. Summary of Aging Management Programs for Structures and Component Supports Evaluated in Chapters II and III of the GALL Report

ID	Type	Component	Aging Effect/Mechanism	Aging Management Programs	Further Evaluation Recommended	Related Generic Item	Unique Item
33	BWR/ PWR	Groups 1-5: concrete	Reduction of strength and modulus due to elevated temperature	Plant-specific	Yes, plant-specific if temperature limits are exceeded	T-10	III.A1-1 III.A2-1 III.A3-1 III.A4-1 III.A5-1
34	BWR/ PWR	Group 6: Concrete; all	Cracking, loss of bond, loss of material due to corrosion of embedded steel; increase in porosity and permeability, cracking, loss of material due to aggressive chemical attack	Inspection of Water-Control Structures Assoc with Nuclear Power Plants and for inaccessible concrete, exam of rep. samples of below-grade concrete, and periodic monitoring of groundwater, if environment is non-aggressive. Plant specific if environment is aggressive.	Yes, plant-specific if environment is aggressive	T-18 T-19	III.A6-1 III.A6-3
35	BWR/ PWR	Group 6: exterior above and below grade concrete foundation	Loss of material (spalling, scaling) and cracking due to freeze-thaw	Inspection of Water-Control Structures Associated with Nuclear Power Plants. Evaluation is needed for plants that are located in moderate to severe weathering conditions (weathering index >100 day-inch/yr) (NUREG-1557).	Yes, for inaccessible areas of plants located in moderate to severe weathering conditions	T-15	III.A6-5

Table 5. Summary of Aging Management Programs for Structures and Component Supports Evaluated in Chapters II and III of the GALL Report

ID	Type	Component	Aging Effect/Mechanism	Aging Management Programs	Further Evaluation Recommended	Related Generic Item	Unique Item
36	BWR/ PWR	Group 6: all accessible/ inaccessible reinforced concrete	Cracking due to expansion/ reaction with aggregates	Accessible areas: Inspection of Water-Control Structures Associated with Nuclear Power Plants. None for inaccessible areas if concrete was constructed in accordance with the recommendations in ACI 201.2R-77.	Yes, if concrete was not constructed as stated for inaccessible areas	T-17	III.A6-2
37	BWR/ PWR	Group 6: exterior above and below grade reinforced concrete foundation interior slab	Increase in porosity and permeability, loss of strength due to leaching of calcium hydroxide	For accessible areas, Inspection of Water-Control Structures Associated with Nuclear Power Plants. None for inaccessible areas if concrete was constructed in accordance with the recommendations in ACI 201.2R-77.	Yes, if concrete was not constructed as stated for inaccessible areas	T-16	III.A6-6
38	BWR/ PWR	Groups 7, 8: Tank liners	Cracking due to stress corrosion cracking; loss of material due to pitting and crevice corrosion	Plant-specific	Yes, plant specific	T-23	III.A7-11 III.A8-9
39	BWR/ PWR	Support members; welds; bolted connections; support anchorage to building structure	Loss of material due to general and pitting corrosion	Structures Monitoring Program	Yes, if not within the scope of the applicant's structures monitoring program	T-30	III.B2-10 III.B3-7 III.B4-10 III.B5-7

Table 5. Summary of Aging Management Programs for Structures and Component Supports Evaluated in Chapters II and III of the GALL Report

ID	Type	Component	Aging Effect/Mechanism	Aging Management Programs	Further Evaluation Recommended	Related Generic Item	Unique Item
40	BWR/ PWR	Building concrete at locations of expansion and grouted anchors; grout pads for support base plates	Reduction in concrete anchor capacity due to local concrete degradation/ service-induced cracking or other concrete aging mechanisms	Structures Monitoring Program	Yes, if not within the scope of the applicant's structures monitoring program	T-29	III.B1.1-1 III.B1.2-1 III.B1.3-1 III.B2-1 III.B3-1 III.B4-1 III.B5-1
41	BWR/ PWR	Vibration isolation elements	Reduction or loss of isolation function/ radiation hardening, temperature, humidity, sustained vibratory loading	Structures Monitoring Program	Yes, if not within the scope of the applicant's structures monitoring program	T-31	III.B4-12
42	BWR/ PWR	Groups B1.1, B1.2, and B1.3: support members: anchor bolts, welds	Cumulative fatigue damage (CLB fatigue analysis exists)	TLAA evaluated in accordance with 10 CFR 54.21(c)	Yes, TLAA	T-26	III.B1.1-12 III.B1.2-9 III.B1.3-9
43	BWR/ PWR	Groups 1-3, 5, 6: all masonry block walls	Cracking due to restraint shrinkage, creep, and aggressive environment	Masonry Wall Program	No	T-12	III.A1-11 III.A2-11 III.A3-11 III.A5-11 III.A6-10
44	BWR/ PWR	Group 6 elastomer seals, gaskets, and moisture barriers	Loss of sealing due to deterioration of seals, gaskets, and moisture barriers (caulking, flashing, and other sealants)	Structures Monitoring Program	No	TP-7	III.A6-12

Table 5. Summary of Aging Management Programs for Structures and Component Supports Evaluated in Chapters II and III of the GALL Report

ID	Type	Component	Aging Effect/Mechanism	Aging Management Programs	Further Evaluation Recommended	Related Generic Item	Unique Item
45	BWR/ PWR	Group 6: exterior above and below grade concrete foundation; interior slab	Loss of material due to abrasion, cavitation	Inspection of Water-Control Structures Associated with Nuclear Power Plants	No	T-20	III.A6-7
46	BWR/ PWR	Group 5: Fuel pool liners	Cracking due to stress corrosion cracking; loss of material due to pitting and crevice corrosion	Water Chemistry and Monitoring of spent fuel pool water level and level of fluid in the leak chase channel.	No	T-14	III.A5-13
47	BWR/ PWR	Group 6: all metal structural members	Loss of material due to general (steel only), pitting and crevice corrosion	Inspection of Water-Control Structures Associated with Nuclear Power Plants. If protective coatings are relied upon to manage aging, protective coating monitoring and maintenance provisions should be included.	No	T-21	III.A6-11
48	BWR/ PWR	Group 6: earthen water control structures - dams, embankments, reservoirs, channels, canals, and ponds	Loss of material, loss of form due to erosion, settlement, sedimentation, frost action, waves, currents, surface runoff, seepage	Inspection of Water-Control Structures Associated with Nuclear Power Plants	No	T-22	III.A6-9
49	BWR	Support members; welds; bolted connections; support anchorage to building structure	Loss of material/ general, pitting, and crevice corrosion	Water Chemistry and ISI (IWF)	No	TP-10	III.B1.1-11

Table 5. Summary of Aging Management Programs for Structures and Component Supports Evaluated in Chapters II and III of the GALL Report

ID	Type	Component	Aging Effect/Mechanism	Aging Management Programs	Further Evaluation Recommended	Related Generic Item	Unique Item
50	BWR/ PWR	Groups B2, and B4: galvanized steel, aluminum, stainless steel support members; welds; bolted connections; support anchorage to building structure	Loss of material due to pitting and crevice corrosion	Structures Monitoring Program	No	TP-6	III.B2-7 III.B4-7
51	BWR/ PWR	Group B1.1: high strength low-alloy bolts	Cracking due to stress corrosion cracking; loss of material due to general corrosion	Bolting Integrity	No	T-27 TP-9	III.B1.1-3 III.B1.1-4
52	BWR/ PWR	Groups B2, and B4: sliding support bearings and sliding support surfaces	Loss of mechanical function due to corrosion, distortion, dirt, overload, fatigue due to vibratory and cyclic thermal loads	Structures Monitoring Program	No	TP-1 TP-2	III.B2-2 III.B4-2 III.B2-3 III.B4-3
53	BWR/ PWR	Groups B1.1, B1.2, and B1.3: support members: welds; bolted connections; support anchorage to building structure	Loss of material due to general and pitting corrosion	ISI (IWF)	No	T-24	III.B1.1-13 III.B1.2-10 III.B1.3-10

Table 5. Summary of Aging Management Programs for Structures and Component Supports Evaluated in Chapters II and III of the GALL Report

ID	Type	Component	Aging Effect/Mechanism	Aging Management Programs	Further Evaluation Recommended	Related Generic Item	Unique Item
54	BWR/ PWR	Groups B1.1, B1.2, and B1.3: Constant and variable load spring hangers; guides; stops	Loss of mechanical function due to corrosion, distortion, dirt, overload, fatigue due to vibratory and cyclic thermal loads	ISI (IWF)	No	T-28	III.B1.1-2 III.B1.2-2 III.B1.3-2
55	PWR	Steel, galvanized steel, and aluminum support members; bolted connections; support anchorage to building structure	Loss of material due to boric acid corrosion	Boric Acid Corrosion	No	T-25 TP-3	III.B1.1-14 III.B1.2-11 III.B2-11 III.B3-8 III.B4-11 III.B5-8 III.B1.1-8 III.B1.2-6 III.B1.3-6 III.B2-6 III.B3-4 III.B4-6 III.B5-4
56	BWR/ PWR	Groups B1.1, B1.2, and B1.3: Sliding surfaces	Loss of mechanical function due to corrosion, distortion, dirt, overload, fatigue due to vibratory and cyclic thermal loads	ISI (IWF)	No	T-32	III.B1.1-5 III.B1.2-3 III.B1.3-3

Table 5. Summary of Aging Management Programs for Structures and Component Supports Evaluated in Chapters II and III of the GALL Report

ID	Type	Component	Aging Effect/Mechanism	Aging Management Programs	Further Evaluation Recommended	Related Generic Item	Unique Item
57	BWR/ PWR	Groups B1.1, B1.2, and B1.3: Vibration isolation elements	Reduction or loss of isolation function/ radiation hardening, temperature, humidity, sustained vibratory loading	ISI (IWF)	No	T-33	III.B1.1-15 III.B1.2-12 III.B1.3-11
58	BWR/ PWR	Galvanized steel and aluminum support members; welds; bolted connections; support anchorage to building structure exposed to air – indoor uncontrolled	None	None	NA - No AEM or AMP	TP-8 TP-11	III.B1.1-6 III.B1.2-4 III.B1.3-4 III.B2-4 III.B3-2 III.B4-4 III.B5-2 III.B1.1-7 III.B1.2-5 III.B1.3-5 III.B2-5 III.B3-3 III.B4-5 III.B5-3

Table 5. Summary of Aging Management Programs for Structures and Component Supports Evaluated in Chapters II and III of the GALL Report

ID	Type	Component	Aging Effect/Mechanism	Aging Management Programs	Further Evaluation Recommended	Related Generic Item	Unique Item
59	BWR/ PWR	Stainless steel support members; welds; bolted connections; support anchorage to building structure	None	None	NA - No AEM or AMP	TP-4 TP-5	III.B1.1-10 III.B1.2-8 III.B1.3-8 III.B2-9 III.B3-6 III.B4-9 III.B5-6 III.B1.1-9 III.B1.2-7 III.B1.3-7 III.B2-8 III.B3-5 III.B4-8 III.B5-5

Table 6. Summary of Aging Management Programs for the Electrical Components Evaluated in Chapter VI of the GALL Report

ID	Type	Component	Aging Effect/Mechanism	Aging Management Programs	Further Evaluation Recommended	Related Generic Item	Unique Item
1	BWR/PWR	Electrical equipment subject to 10 CFR 50.49 environmental qualification (EQ) requirements	Degradation due to various aging mechanisms	Environmental qualification of electric components	Yes, TLAA	L-05	VI.B-1
2	BWR/PWR	Electrical cables, connections and fuse holders (insulation) not subject to 10 CFR 50.49 EQ requirements	Reduced insulation resistance and electrical failure due to various physical, thermal, radiolytic, photolytic, and chemical mechanisms	Electrical cables and connections not subject to 10 CFR 50.49 EQ requirements	No	L-01 LP-03	VI.A-2 VI.A-6
3	BWR/PWR	Conductor insulation for electrical cables and connections used in instrumentation circuits not subject to 10 CFR 50.49 EQ requirements that are sensitive to reduction in conductor insulation resistance (IR)	Reduced insulation resistance and electrical failure due to various physical, thermal, radiolytic, photolytic, and chemical mechanisms	Electrical Cables And Connections Used In Instrumentation Circuits Not Subject To 10 CFR 50.49 EQ Requirements	No	L-02	VI.A-3
4	BWR/PWR	Conductor insulation for inaccessible medium voltage (2 kV to 35 kV) cables (e.g., installed in conduit or direct buried) not subject to 10 CFR 50.49 EQ requirements	Localized damage and breakdown of insulation leading to electrical failure due to moisture intrusion, water trees	Inaccessible medium voltage cables not subject to 10 CFR 50.49 EQ requirements	No	L-03	VI.A-4

Table 6. Summary of Aging Management Programs for the Electrical Components Evaluated in Chapter VI of the GALL Report

ID	Type	Component	Aging Effect/Mechanism	Aging Management Programs	Further Evaluation Recommended	Related Generic Item	Unique Item
5	PWR	Connector contacts for electrical connectors exposed to borated water leakage	Corrosion of connector contact surfaces due to intrusion of borated water	Boric Acid Corrosion	No	L-04	VI.A-5
6	BWR/PWR	Fuse Holders (Not Part of a Larger Assembly): Fuse holders – metallic clamp	Fatigue due to ohmic heating, thermal cycling, electrical transients, frequent manipulation, vibration, chemical contamination, corrosion, and oxidation	Fuse holders	No	LP-01	VI.A-8
7	BWR/PWR	Metal enclosed bus – Bus/connections	Loosening of bolted connections due to thermal cycling and ohmic heating	Metal Enclosed Bus	No	LP-04	VI.A-11
8	BWR/PWR	Metal enclosed bus – Insulation/insulators	Reduced insulation resistance and electrical failure due to various physical, thermal, radiolytic, photolytic, and chemical mechanisms	Metal Enclosed Bus	No	LP-05	VI.A-14
9	BWR/PWR	Metal enclosed bus – Enclosure assemblies	Loss of material due to general corrosion	Structures Monitoring Program	No	LP-06	VI.A-13
10	BWR/PWR	Metal enclosed bus – Enclosure assemblies	Hardening and loss of strength due to elastomers degradation	Structures Monitoring Program	No	LP-10	VI.A-12

Table 6. Summary of Aging Management Programs for the Electrical Components Evaluated in Chapter VI of the GALL Report

ID	Type	Component	Aging Effect/Mechanism	Aging Management Programs	Further Evaluation Recommended	Related Generic Item	Unique Item
11	BWR/ PWR	High voltage insulators	Degradation of insulation quality due to presence of any salt deposits and surface contamination; Loss of material caused by mechanical wear due to wind blowing on transmission conductors	Plant specific	Yes, plant specific	LP-07 LP-11	VI.A-9 VI.A-10
12	BWR/ PWR	Transmission conductors and connections; switchyard bus and connections	Loss of material due to wind induced abrasion and fatigue; loss of conductor strength due to corrosion; increased resistance of connection due to oxidation or loss of preload	Plant specific	Yes, plant specific	LP-08 LP-09	VI.A-16 VI.A-15
13	BWR/ PWR	Cable Connections – Metallic parts	Loosening of bolted connections due to thermal cycling, ohmic heating, electrical transients, vibration, chemical contamination, corrosion, and oxidation	Electrical cable connections not subject to 10 CFR 50.49 environmental qualification requirements	No	LP-12	VI.A-1
14	BWR/ PWR	Fuse Holders (Not Part of a Larger Assembly) Insulation material	None	None	NA - No AEM or AMP	LP-02	VI.A-7

APPENDIX

LISTING OF PLANT SYSTEMS
EVALUATED IN THE GALL REPORT
(VOLUME 2)

A-2

Plant Systems
Evaluated in the GALL Report (Volume 2)

Type	System	Section in GALL (Vol. 2)
BWR	Automatic depressurization system	V.D2
BWR	Containment structures:	
	Mark I steel containments	II.B1
	Mark II concrete and steel containments	II.B2
	Mark III concrete and steel containments	II.B3
	Common components	II.B4
BWR	High-pressure coolant injection	V.D2
BWR	High-pressure core spray	V.D2
BWR	Low-pressure coolant injection and residual heat removal	V.D2
BWR	Low-pressure core spray	V.D2
BWR	Reactor building	III.A1
BWR	Reactor building with steel superstructure	III.A2
BWR	Reactor coolant pressure boundary	IV.C1
BWR	Reactor coolant system connected systems (up to and including the second isolation valve):	
	Automatic depressurization system	IV.C1
	Feedwater	IV.C1
	High-pressure core spray	IV.C1
	High-pressure coolant injection	IV.C1
	Isolation condenser	IV.C1
	Low-pressure coolant injection	IV.C1
	Low-pressure core spray	IV.C1
	Main steam	IV.C1
	Reactor core isolation cooling	IV.C1
	Reactor water cleanup	IV.C1
	Recirculation system	IV.C1
	Residual heat removal	IV.C1
	Shutdown cooling	IV.C1
	Standby liquid control	IV.C1
BWR	Reactor core isolation cooling	V.D2
BWR	Reactor vessel	IV.A1
BWR	Reactor vessel internals	IV.B1
BWR	Reactor water cleanup system	VII.E3
BWR	Shutdown cooling system (older plants)	VII.E4
BWR	Standby gas treatment system	V.B
BWR	Standby liquid control system	VII.E2
BWR	Suppression pool cleanup system	VII.A5
BWR	Unit vent stack	III.A9
BWR/ PWR	Auxiliary and radwaste area ventilation system	VII.F2
BWR/ PWR	Auxiliary building, diesel generator building, radwaste building, turbine building, switchgear room, auxiliary feedwater pump house, and utility/piping tunnels	III.A3
BWR/ PWR	Carbon steel components	V.E, VII.I, VIII.H
BWR/ PWR	Closed-cycle cooling water system (reactor auxiliary cooling water)	VII.C2

Type	System	Section in GALL (Vol. 2)
BWR/ PWR	Component supports	III.B
BWR/ PWR	Compressed air system	VII.D
BWR/ PWR	Concrete tanks	III.A7
BWR/ PWR	Condensate system	VIII.E
BWR/ PWR	Containment internal structures, excluding refueling canal	III.A4
BWR/ PWR	Containment isolation components (containment isolation valves for in-scope systems are addressed in chapters IV, VII, and VIII)	V.C
BWR/ PWR	Control room/building	III.A1
BWR/ PWR	Control room area ventilation system	VII.F1
BWR/ PWR	Demineralized water makeup	Not in scope of 10 CFR 50.54
BWR/ PWR	Diesel fuel oil system	VII.H1
BWR/ PWR	Diesel generator building ventilation system	VII.F4
BWR/ PWR	Electrical components	VI.A, B
BWR/ PWR	Emergency diesel generator system	VII.H2
BWR/ PWR	Extraction steam system	VIII.C
BWR/ PWR	Feedwater system	VIII.D2, D1
BWR/ PWR	Fire protection	VII.G
BWR/ PWR	Fuel storage facility and refueling canal	III.A5
BWR/ PWR	Heating and ventilation systems	VII.F1, F2, F3, F4
BWR/ PWR	Main steam system	VIII.B2, B1
BWR/ PWR	New and spent fuel storage	VII.A1, A2
BWR/ PWR	Open-cycle cooling water system (service water system)	VII.C1
BWR/ PWR	Overhead heavy load and light load (related to refueling) handling systems	VII.B
BWR/ PWR	Potable and sanitary water	Not in scope of 10 CFR 50.54
BWR/ PWR	Primary containment heating and ventilation system	VII.F3
BWR/ PWR	Refueling canal	III.A5
BWR/ PWR	Spent fuel pool cooling and cleanup	VII.A3, A4
BWR/ PWR	Steam turbine system	VIII.A
BWR/ PWR	Steel tanks	III.A8
BWR/ PWR	Ultimate heat sink	VII.C3
BWR/ PWR	Water-control structures (e.g., intake structure, cooling tower, and spray pond)	III.A6
PWR	Accumulators	V.D1
PWR	Auxiliary feedwater system	VIII.G
PWR	Chemical and volume control system	VII.E1
PWR	Combustible gas control (containment H_2 control)	V.E1
PWR	Containment spray system	V.A
PWR	Containments: Concrete containments Steel containments Common components	II.A1 II.A2 II.A3
PWR	Coolant storage/refueling water system	V.D1

Plant Systems
Evaluated in the GALL Report (Volume 2) (continued)

Type	System	Section in GALL (Vol. 2)
PWR	Core flood system (see accumulators or safety injection tanks)	V.D1
PWR	High-pressure safety injection	V.D1
PWR	Lines to chemical and volume control system	V.D1
PWR	Low-pressure safety injection	V.D1
PWR	Shield building	III.A1
PWR	Reactor coolant system and connected lines (up to and including the second isolation valve): 　Chemical and volume control system 　Core flood system 　Drains and instrumentation lines 　High-pressure injection system 　Low-pressure injection 　Residual heat removal or shutdown cooling 　Safety injection 　Sampling system	 IV.C2 IV.C2 IV.C2 IV.C2 IV.C2 IV.C2 IV.C2 IV.C2
PWR	Reactor coolant system, pressurizer, pressurizer relief tank, and other Class 1 components	IV.C2
PWR	Reactor vessel	IV.A2
PWR	Reactor vessel internals	IV.B2, B3, B4
PWR	Residual heat removal or shutdown cooling	V.D1
PWR	Safety injection tanks	V.D1
PWR	Steam generator blowdown system	VIII.F
PWR	Steam generators	IV.D1, D2

www.ingramcontent.com/pod-product-compliance
Lightning Source LLC
Chambersburg PA
CBHW081503170526
45166CB00008B/2533